By Vitus B. Dröscher

THE MYSTERIOUS SENSES OF ANIMALS

THE MAGIC OF THE SENSES

THE FRIENDLY BEAST

THE FRIENDLY BEAST
Latest Discoveries in Animal Behavior

Translated from the German by
Richard and Clara Winston

THE FRIENDLY BEAST

BEAST

Latest Discoveries in Animal Behavior

by
VITUS B. DRÖSCHER

E. P. DUTTON & CO., INC.

New York 1971

First published in the U.S.A. 1971 by E. P. Dutton & Co., Inc.
Die freundliche Bestie © Gerhard Stalling Verlag, 1968
English translation Copyright © 1970 by E. P. Dutton & Co., Inc.
All rights reserved. Printed in the U.S.A.

FIRST EDITION

Published simultaneously in Canada by Clarke, Irwin & Company Limited, Toronto and Vancouver

Library of Congress Catalog Card Number: 76-95487

SBN 0-525-109854

Line drawings in this book are by Helmut Skarupp, after originals in the following sources: Page 9: *Science* 158 (1967) and Theodore Dobzhansky, *Die Entwicklung zum Menschen* (Hamburg: Paul Parey Verlag, 1958). P. 19: Jane van Lawick-Goodall, *My Friends the Chimpanzees* (Washington, D.C.: The National Geographic Society, 1967). Pp. 29, 30: *Science* 150 (1965). Pp. 58, 59: Jean Rivolier: *Gast bei den Pinguinen* (Stuttgart, Hans E. Gunther Verlag, 1963). Pp. 62, 66, 67: *Zeitschrift für Tierpsychologie* 21 (1964). P. 76: *Scientific American* 201 (1959). P. 81: *Naturwissenschaftliche Rundschau* 21 (1968). P. 83: *Zeitschrift für Tierpsychologie* 24 (1967). P. 97: *Scientific American* 211 (1964). P. 132: Charles Darwin, *The Expression of Emotions in Man and Animals* (London, 1872). P. 164: Detley Ploog, *The Behavior of Squirrel Monkeys* (Chicago: University of Chicago Press, 1965). P. 170: Bernhard Grzimek, ed.: *Grzimeks Tierleben* (Munich: Kindler Verlag, 1967). P. 176: *Scientific American* 100 (1959). Pp. 180, 182, 193, 203: Bernhard Grzimek, ed.: *Grzimeks Tierleben* (Munich: Kindler Verlag, 1967). P. 218: *Scientific American* 209 (1963).

Contents

Acknowledgments

For their encouragement, advice, and careful reading of the German manuscript the author and his publisher would like to thank Dr. Michael Abs, Dr. Otto von Frisch, Mr. Sigrid Hopf, Dr. Adriaan Kortlandt, Dr. Hubert Markl, Professor Günther Niethammer, Professor Detlev Ploog, Professor K. Poeck, Dr. Thomas Schultze-Westrum, and Professor Erwin Stresemann.

Picture Credits

Toni Angermayer (Plates 26, 27); Anthony-Verlag (Plates 15, 16); Des Bartlett, Bruce Coleman Ltd. (Plate 29); Bavaria-Verlag (Plates 2, 14, 19, 31); Central Press Photos Ltd. (Plate 23); dpa (Plates 4, 5, 12); Herbert Grenzemann (Plate 7); Sigrid Hopf, aus D. Ploog; 'Verhaltensforschung als Grundlagenwissenschaft für die Psychiatrie' (Plate 28); Dr. Adriaan Kortlandt (Plate 3); Tierbilder Okapia (Plates 18, 20); P.-A. Reuters Photos Ltd. (Plate 30); Greta Robok (Plates 8, 10, 25, 34); Jürgen Schmidt (Plate 35); Fritz Siedel (Plate 13); Heinz Sielmann, aus d. Film 'Galapagos' (Plate 32); Süddeutscher Verlag (Plates 11, 21, 24, 33); United Press International Photo (Plate 6); Lies Wiegman (Plate 24); Walter Wissenbach (Plates 9, 17); J. C. J. van Zon (Plate 1).

THE FRIENDLY BEAST
Latest Discoveries in Animal Behavior

1. Arrested Human Beings

Chimpanzees Fight with Clubs

August 3, 1965, in Central Africa. Halfway up the trunk of a huge tree a shapeless mass of leaves bulges. Two green lianas descend unobtrusively from this bulge. In reality they form a rope ladder. In the cover of the leaves, forty feet above ground, two men have been crouching daily for the past seven weeks inside their camouflaged hut of steel posts and plywood panels. They are Dr. Adriaan Kortlandt of Amsterdam and Ernest G. Bresser, his cameraman.

Far below them on the edge of the jungle a chimpanzee path winds along. All around is deep silence. The cool morning mist still lingers over the clearing in the forest. Suddenly the men start, and the scientist begins to count silently: Eight . . . eleven . . . fifteen . . . twenty chimpanzees! One after another they come along the path, noiselessly, cautiously peering in all directions, for a deadly enemy may be lurking behind every bush, a man or perhaps a leopard. Most of the chimps are females. Some of them are carrying their babies on their backs, others have infants clinging to their stomachs. Older offspring scamper about freely.

At last, after seven weeks of patient waiting, the scientists have been rewarded by the sight of a sizable troop of chimpanzees. The moment of decision has come. Dr. Kortlandt pulls a rope. A camouflaged trapdoor close to the apes opens, and a stuffed leopard springs from the bushes. Between its paws the beast holds a chimpanzee doll as its 'prey'. As if struck by lightning, the apes stare at the predator. A full thirty seconds.

Then hell breaks loose. Screeching loudly, some of the chimpanzees leap into the surrounding papaya trees. Dangling from the branches, they swing wildly through the air. Their screams echo through the jungle for miles.

The Amsterdam zoologist watches with bated breath. Meanwhile the chimpanzees have accomplished nothing by their wild uproar. Naturally all the noise has no daunting effect on a stuffed leopard. The second phase of their defensive battle now begins. The most courageous of them venture forward a little. Do they realize that their enemy is a fake? The scientist pulls another control. At this the leopard waves his tail, and a clever contraption, actually the motor of a windshield wiper, causes the head to move from side to side. But the apes do not leap back. On the contrary, something quite remarkable happens. The chimpanzees stare at the big cat with the utmost resolution, simultaneously signing to their children to remain behind. They make precisely the same sort of hand signals that men would use in similar circumstances. Then the mothers snatch up four-foot clubs, which have been deliberately left lying around, and swing them threateningly back and forth. One after the other, they rush forward with piercing, barking cries. But a few yards in front of the leopard the chimpanzee females stop, hurl the clubs with all their might at the enemy, and run back to their children as if pursued by the Furies. Passionately, they clutch their young to their breasts, then put them down again and rush toward the enemy armed with fresh clubs.

Other apes throw sticks and fruit, pound the trunks of the trees, run up in feigned attacks with branches in hand, or swing uprooted saplings like whips against the enemy.

The chimpanzee attack startlingly resembles the behavior of a Zulu warrior swinging a *kirri*. The ape picks up the club in one hand like a man, straightens up—leaning forward somewhat—and rushes forward on two legs. Obviously the erect gait affords him advantages in combat—that same erect gait that early biologists posited as the distinguishing characteristic of man.

The leopard's immobility gradually seems to arouse doubts in the horde of angry apes. The pauses between advances grow longer. Finally the chimps, who have met with no proper response from the animal, sit down in a semicircle around the decoy, stare curiously at it, scratch their foreheads and arms in perplexity—and now and then gobble down a banana. Then they attack once more, though with considerably abated violence. After about thirty-five minutes they gradually drift away from the battlefield and vanish in the jungle thickets. Twelve uprooted saplings or broken branches remain as evidence of the battle.

The significance of this experiment is that it proves that animals, not only man, use weapons in battle—and, moreover, in the erect posture. Or is it a mistake to regard chimpanzees as animals?

Erect posture and the use of weapons—here we truly have a unique combination. In itself walking on two legs brings only disadvantages in the struggle for survival. Dr. Kortlandt (whose account is written in English), concluded: 'When fleeing from a predator, or from any other dangerous animal (great ape, elephant, buffalo, etc.), a biped is much more hindered by the tangle of vegetation and remains visible to the enemy in his full length. When brought to a stand, the vulnerable parts are fully exposed and the throat is in the best position a carnivore could wish.

'Nearly all quadrupedal animals lower their head and shoulders when fighting or when defending themselves. There are some exceptions: bears, stallions and deer bucks (in the season when they have no antlers) may fight on their hind legs and use their front legs to beat the opponent, but in these cases their claws and hoofs are tremendous weapons. Unarmed, man has nothing of this kind. It seems absolutely unbelievable that bipedal hominids could ever have survived in the wild without having reasonably effective weapons.'[1]*

On the other hand, imagine a lion picking up a club. It is utterly

*The superior figures refer to the references in the Source Notes, beginning on page 227.

impossible. Seizing a weapon with one hand is possible only for the biped. Furthermore, upright posture offers advantages in the presence of the enemy only if some kind of weapon is available. Even so, the creature fighting on two legs is in a dangerously exposed position. That is evident from the terror and courage which the stuffed leopard aroused in the chimpanzees; it is likewise evident in the conduct of a troop of Masai warriors of the East African plains who hunt a solitary lion with spears.

Nevertheless, only this kind of fighting could have permitted an anthropoid-like creature, in remote primordial times, to develop into that muscularly weak, slow-running, unarmored creature, an inept climber with a harmlessly small mouth, whom we call *Homo Sapiens*.

Throwing weapons is instinctive with chimpanzees, but manipulating weapons has to be learned. Dr. Kortlandt[2, 3, 4] was able to demonstrate that impressively by an experiment in the Rotterdam Zoo. In one enclosure was a chimpanzee that had been born in the zoo and had never in its life seen a large predator, let alone thrown objects at one. First the zoologist gave it a dozen wooden blocks. As expected, it paid little attention to them. Then a rather playful adolescent tiger was released in an adjacent cage. The chimpanzee's explosion of rage was something to see. Instantly it began hurling the wooden blocks in all directions against the bars of its cage, at the same time letting forth howls that terrified the tiger.

Throwing in itself, therefore, is an innate reaction to fend off predatory enemies. But aiming the objects thrown, like fighting with clubs, is a skill to be learned.

If both man and his nearest relative in the animal kingdom, the chimpanzee, use weapons in time of danger, it seems a likely premise that at least the disposition to do so must have already existed in that creature which was the common ancestor of man and chimpanzee. 'Therefore the evolution, which today has reached the phase of intercontinental atomic rockets, must have begun some ten or fifteen million years ago,' Adriaan Kortlandt remarks.[5] In this book we shall examine the roots of human evolution on the basis of the latest results of scientific research.

If an Ape Had Invented the Spear . . .

There is one factor in the course of that battle with clubs against the leopard that does not quite satisfy us. Despite all the commotion, not a single chimpanzee actually hit the stuffed leopard; none of the thrown weapons so much as glanced off it, and no direct blows with the clubs were delivered. The same thing happened in five further experiments conducted by Dr. Kortlandt in 1963 and 1964 with different groups of chimpanzees in the jungles of the Congo. It seemed highly doubtful, therefore, whether his behavior could be called genuine employment of weapons.

Sometimes, however, crucial insights can be obtained from such disappointments when all the circumstances are thoroughly analyzed. The fact remains that chimpanzees, whether born in the zoo or in the jungle, employ long sticks for gesticulating, striking, and throwing. In dense jungle underbrush, however, that is not a very useful form of defense. The sticks are constantly being caught by branches, vines, and shrubs. The described experiment, as a matter of fact, was carried out under atypical conditions, in a small clearing, so that the 'battle' could be conveniently watched and filmed.

Because of these objections it seemed scarcely credible that the observed technique of employing weapons could have arisen in dense jungle. But what if chimpanzees in former ages did not live in the jungle, but on the plains and savannas?

Around the year 1500, when America was just being discovered, large parts of South Africa were not yet settled or only thinly populated by human beings. At that time a tremendous human migration began on the black continent. Warlike Bantu tribes[6] pressed forward in great numbers from north to south. They drove Pygmies, the original inhabitants of the area they conquered, into the jungles of the Congo. Under the resultant pressure the equally dwarflike Bushmen sought refuge in the semidesert of Kalahari in southern Africa. Probably the hordes of chimpanzees were similarly driven by men from the open country into the forests.

In the jungles the chimpanzee's talent for using weapons must have atrophied for lack of use. With the practiced eye of the animal psychologist, Adriaan Kortlandt at once recognized that the animals were handling their clubs clumsily, as if they had scarcely ever practiced using them. Moreover, Professor Wolfgang Köhler[7] had already described in 1917 how newly captured jungle chimpanzees initially cannot throw well or strike accurately with clubs, but gradually learn both techniques in open enclosures. Kortlandt wondered how chimpanzees that live in the savannas rather than the jungles would behave in the same situation.

A great deal depends on the answer to this question. It can serve as an important clue to the origin of our ancestors. If savanna chimpanzees behave no differently from their fellows of the jungle, the question would have to remain unanswered. But if they can use weapons more skillfully, in a more manlike manner, so to speak, there would be a clear sign that the common ancestor of chimpanzee and man did not 'crouch in trees, hairy and fierce of face,' as Erich Kästner puts it, but was at least to some extent a dweller in open country.

Fortunately, there are still some savanna chimpanzees living today. Significantly, they dwell only in places where they are not hunted by men. In Guinea the religion of the Mohammedan population forbids killing apes or eating their flesh. Accordingly, three of Dr. Kortlandt's colleagues, doctoral candidates Jo van Orshoven, Hans van Zon, and Rein Pfeyffers, organized the group's sixth chimpanzee expedition, of 1966–67, in that West African country.

Chimpanzees of the savanna establish regular lanes through their territory. Their paths differ from those of men only in sometimes passing under low-hanging boughs which men would avoid. The apes even tramp the soil hard on their paths[8] to keep down the growth of weeds in which poisonous snakes might lurk. If crossing through high grass becomes unavoidable, they run stiff-legged, like a city man in the country after someone has shouted at him: 'Watch out for the rattler!'

Along one such path Dr. Kortlandt's associates[9, 10] set up a well-camouflaged camera observation post, and concealed the stuffed leopard under grass and twigs a little in front of the camera. After they had waited some weeks, a chimpanzee troop of between fifteen and eighteen animals came along. The scientists released the stuffed predator from its "ambush". What followed has been recorded by the scientists in a unique documentary film.

With a piercing scream, the chimps took several jumps backward. They jerked small trees out of the ground, or picked up sticks lying about. Thus armed, they ran forward on two legs, in erect posture, and formed a semicircle around the leopard. Groups of two to five males and females, some of the latter with babies clinging to their backs or bellies, took turns dashing at it. Precisely aimed blows of the clubs rained down on the decoy. It was calculated from the films that some of the blows crashed down on the stuffed animal at a speed of sixty miles an hour. A real leopard would probably have been driven to flight in a few minutes, or else it would quickly have been as dead as the stuffed pelt.

The chimpanzees then surrounded the predator. The bravest of them seized the enemy by the tail and dragged him some seventy feet into the bush. In the course of this operation the head broke from the body.

This really settled the matter. But although the leopard's detached head lay on the grass at a great distance from the body, it exerted a mysterious suggestive force upon the chimpanzees. They continued to keep a respectful distance from it, as if the head were a complete, still living, and dangerous predator. From time to time a warrior would run up and deliver a ferocious blow of a stick upon the head. Was this strange effect due to the decoy's glass eyes, which still stared unwinkingly at the chimpanzees? At any rate the troop of chimps did not calm down until dusk settled. Then they vanished in the darkness of the night.

African game wardens have frequently reported that leopards meeting a troop of chimpanzees by daylight will hurriedly get out of the way. Now we know why. The leopards have their chance of victory only at night, when they can fall on the apes in their sleep.

Above all, however, the dramatic battle with clubs shows what must have happened long ago when chimpanzee-like animals first encountered their strongest rival for food, man's ancestor. This was a creature who stood in the evolutionary scale between the manlike apes (Australopitheci) and the apelike men (Pithecanthropi).[11] Probably a pre-hominid armed with a club had not much more chance against a pre-chimpanzee armed with a club than would a man of today. Zoologists recommend never tangling with a full-grown chimpanzee. The playful, entertaining chimpanzees shown us in circus shows are really only small children—not even adolescents. A full-grown chimpanzee on the other hand has about twice the muscular strength of an average man. His teeth are scarcely inferior to the leopard's; the fierceness of his offensive leaps and his lightning reactions are far superior to men's.

Pre-man would no doubt have had a very difficult time with such an enemy. To this day the Pygmies refer to fighting with apes not as 'hunting' but as 'war'. Probably, the Dutch scientist conjectures, a significant improvement in weapons and battle technique must have been necessary for the issue in that primordial struggle to be decided in favor of pre-man, and thus of man. The decisive development, he thinks, must have been the transformation of the club into a spear, in other words, the close-quarters weapon into a weapon that could be used in open country for stabbing, throwing, and killing from a safe distance.

Once man or pre-man had such a weapon, the ancestors of the chimpanzees were driven from their homes in many places, to lead fugitive lives in the jungles. This event must have had a paralyzing effect upon the evolution of the chimpanzees. In South Africa and in the East African Olduvai Gorge, Professor R. A. Dart[12] and Dr. L. S. B. Leakey[13] have found the fossilized bone fragments of many manlike apes that peopled the earth from about two million years ago to about four hundred thousand years ago. From the size of the skulls we can conclude that the brain of these pre-hominids was scarcely larger than that of present-day chimpanzees. Nevertheless they were able to develop a humanoid type of culture.[14] It has been shown that they worked antelope bones into weapons with which they could kill animals.

Our nearest relative in the animal kingdom, however, remained an animal. For after man had come into being, human characteristics no

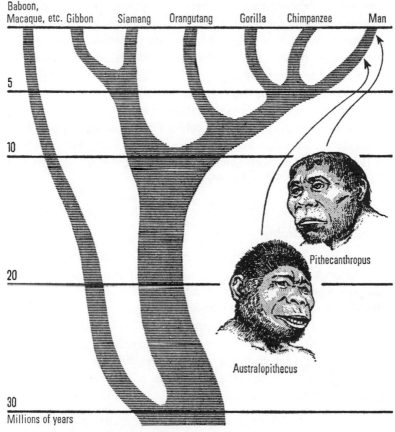

Baboon,
Macaque, etc. Gibbon   Siamang   Orangutang   Gorilla   Chimpanzee   Man

Pithecanthropus

Australopithecus

Millions of years

The common ancestry of man and ape. (After V. M. Sarich and A. C. Wilson.)

longer had any value for chimpanzee survival. The chimpanzee's intelligence was checked by life in the jungle; he remained a kind of arrested human being.

Worse still: judged by human standards, the chimpanzee evidently evolved regressively in intelligence. Probably the common ancestor of man and chimpanzee was considerably more manlike than any anthropoid ape of the present day. This is the burden of Dr. Kortlandt's hypothesis of dehumanization, a hypothesis given significant support by the experiments described above.

Jestingly, we might ask now: what would the world probably look like today if, in those primordial times, the pre-chimpanzee rather than the pre-man had invented the spear? Here is certainly a rich subject for authors of science fiction. But it was certainly not pure chance that this intellectual lightning flashed in the mind of a human ancestor. For after all, the pre-chimpanzee was not capable of constructing a spear. Why not is a matter that will concern us again in the last chapter of this book. For the present let us track down other roots of the characteristics that may be regarded as 'typically human'.

Animals in Mourning

One day, after another chimpanzee club battle in the Congo jungle, Dr. Kortlandt discovered further evidence of how many hitherto unsuspected human traits can be hidden within animals. He witnessed animals mourning for a dead fellow.

After the half-hour uproar around the stuffed leopard that held a chimpanzee doll in its paws, the apes retired to their nocturnal sleeping places. That night the zoologist removed the stuffed leopard, but left the chimpanzee doll lying at the 'scene of the crime'. He describes what happened next morning as follows:[15]

At the first light of dawn the chimpanzee troop returned. In funereal silence they all assembled in a wide circle around the doll. Slowly, a few of them ventured closer. Finally, a mother with her baby clinging to her abdomen stepped forward out of the silent circle. Cautiously, she approached the 'victim' and sniffed at it. Then she turned to the assembled horde and shook her head. Thereafter each ape silently departed. Only one chimpanzee crippled by polio (strangely, these apes also suffer severely from infantile paralysis) remained for a while sitting beside the 'corpse', looking steadily at it. It was as though he could not take leave of the face of death.

Finally he too went away. After that there was a sustained silence.

All morning long we did not hear a single chimpanzee cry, nor did we later in the day.

But the high point of my whole expedition had been that chimpanzee female's shaking her head after gazing at the dead body. Of course we do not know what the animal wished to communicate to the silent onlookers. Perhaps it meant: 'No, unfortunately no sign of life.' But more probably it was: 'No, not any one of us.' We are equally ignorant of why that somber mood descended upon all the members of the troop. But an unexpected world always lies hidden behind these chimpanzee faces.

This memorable incident gave the scientist an idea. He decided to find out how apes in the wild would react to a picture of a chimpanzee. He therefore placed on the jungle path a zoo poster showing a chimpanzee's head in close-up.

As soon as the first animal saw the photo it stood still in horror. All the other chimpanzees likewise reacted with fear of this picture of one of their own kind, a far greater fear than they had of leopards and poisonous snakes. Finally, they actually fled, into a bypath. Only the ape whose left arm was paralyzed from polio was attracted to the 'face of death' here also. For a long time he looked mutely at the picture, scratching his head.

The animals reacted to a very small chimpanzee doll by flight, and also to their mirror image in the rearview mirror that Adriaan Kortlandt had removed from his cross-country vehicle.

A curious fact may throw light on this behavior. As long as thirty or forty thousand years ago human beings began painting realistic pictures of animals on the rock of their caves. But they ventured naturalistic images of themselves only during the last few thousand years. Did this taboo spring from some similar primeval fear of their own image?

These chimpanzee reactions are, at any rate, unique in the entire animal kingdom. In general animals have no fear of dead things, consequently none of lifeless images. A stag passes indifferently by the cadaver of a member of its own herd. Former friends or enemies will sniff briefly at a dead dog and then let it lie without any sign of emotion. To be sure, occasionally a parakeet will mourn its dead cagemate or a dog will break down after its master's death. But this is sorrow for individual loss, in no way comparable to the dread shown by the chimpanzees.

The American scientist Dr. R. A. Butler[16] has conducted a somewhat grim experiment with rhesus monkeys. He cut off the head of a dead animal, set up the body against the bars of the cage, and placed the head in the hands, which he positioned over the abdomen. The other rhesus monkeys seemed wholly unaffected by this ghastly sight.

Chimpanzees, on the other hand, react with great terror to the sight of dead members of their species and even other dead mammals (except those they have killed themselves). A severed chimpanzee arm or leg is sufficient to send a chimpanzee running away frantically. They even fear sleeping animals.

For this reason Dr. Kortlandt believes that chimpanzees definitely have an inkling of the meaning of death.

In all probability there is a connection between this capacity and the chimpanzee's need, unusual in animals, to rescue members of the horde at the risk of its own life. Long ago Professor Wolfgang Köhler[17] made some highly interesting observations during a safari.

A chimpanzee was wounded by one of the hunters and fell to the ground. When it uttered a shrill cry for help, the other members of the horde surrounded it, raised it up, supported it with 'incredibly human gestures,' and mouthing gentle sounds urged it to walk. Meanwhile a powerful ape rushed forward and interposed itself between the hunters and the wounded chimpanzee with its helpers. It fled to safety only after hearing repeated calls from its companions, indicating that they were safe in the dense woods.

Such self-sacrificing rescue actions throw a spotlight upon a hitherto unexplained phenomenon: chimpanzees are the only animals that cannot be caught in traps. It had always been said that they were too intelligent to run into traps. Dr. Kortlandt, however, found indications that a trapped chimpanzee is always immediately freed by his companions.

It is particularly amazing that chimpanzees manifest helpfulness not only to their own kind. When the Dutch scientist tied a chick along the jungle path, the robust chimpanzees freed the delicate, peeping little creature from its bonds, and worked so carefully that they did not injure the fragile legs of the small bundle of down. This surprising discovery of affection for animals in an animal strengthened Dr. Kortlandt's suspicion that an old wives' tale among the African natives might be true. Could it be that chimpanzees really did steal human babies?

Naturally a human baby could not be used in an experiment to check the truth of the matter. But there was no reason for not using a monkey baby as a substitute. The Dutch scientist therefore tied a mangabey baby along the chimpanzee path. When a troop came along, all the apes gathered inquistively around the tiny, screeching creature. Finally a young and still childless female tried to untie and bite through the rope—exerting just as much delicacy and care as had been used in freeing the chick. Obviously she wanted to free the baby and perhaps to take it with her. Unfortunately the rope was too strong, the weaker liana with which Kortlandt had intended to tie the monkey having been lost.

For years game wardens in Uganda have observed a troop of chimpanzees which included a long-tailed monkey. There could be only one explanation for this mixed company: the long-tailed monkey must have been picked up by the apes as a baby, abducted, and raised by them—as a kind of pet. Romulus and Remus in the African jungle!

Probably curiosity and playfulness impel the chimpanzees to occupy themselves with other animals. Once they have grown used to each other, none of the chimps seem to mind that the 'foundling' remains much smaller and weaker than the other members of the society, besides looking so different. The 'pet' relationship depends, of course, on at least one chimpanzee female's being in the proper mood to vent her unfulfilled maternal instinct on a surrogate, in this case an alien baby. Apparently the origins of keeping pets must be looked for in some such configuration.

Why, then, should not chimpanzees also abduct a human baby whom its mother has laid down at the edge of the forest during plantation work? For human children, however, such apish love is inevitably fatal. They cannot, like monkey babies, cling to the female's hide, and will fall as soon as the chimpanzee first climbs a tree.

Early in December 1967 it was reported in Johannesburg, South Africa, that a two-year-old girl had been killed by baboons. The parents had driven into the plains for an outing. Stopping their car for a picnic, they had not kept close watch on the child. When they were about to leave, they found that she had vanished. Two days later a party of searchers found the body three miles from the picnic site, with bites inflicted by baboons' teeth!

The distance suggested that the girl had initially been 'adopted' by the apes with no bad intentions and been carried about for some time. But in baboon society even a foundling must be able to behave like a baboon if there are to be no fatal misunderstandings. The child might, for example, have fallen out with the other baboon babies. Young apes are given to teasing and bickering. Perhaps the human child had struck out against them, and the ape mothers had come rushing up in response to cries from the young baboons.

In such a situation only one kind of behavior might have helped: the little girl should have crouched and made a placating gesture signifying: 'Forgive me, I didn't mean to be nasty.' Among baboons this is done by turning back and backside toward the adversary, crouching on the ground like a Moslem at prayer, looking at the threatening baboon over one shoulder, grinning broadly, and smacking the lips audibly. If this ritual is observed, the aggressor turns away, satisfied. The 'apology' is accepted.

The American professors S. L. Washburn and Irven DeVore[18] have effectively tested this formula on the Amboseli Plain of East Africa. It was a highly risky experiment, since the two anthropologists had no way of knowing whether the baboons would accept the 'apology' from human beings. But it worked.

The little girl from Johannesburg naturally knew nothing about animal psychology and did not understand how to counteract aggression on the part of the baboons. That was fatal. On the other hand, the young long-tailed monkey which was taken into chimpanzee society probably succeeded in communicating.

Alone Among Wild Chimpanzees

In any case it is a hazardous feat for a human being to join chimpanzees living in the wild. In July 1960 a twenty-nine-year-old woman, a British student of zoology, ventured it. In spite of many warnings she went to Africa accompanied only by her mother and entered the Gombe Chimpanzee Reservation on the eastern shore of Lake Tanganyika. Beforehand, to be sure, she had thoroughly studied chimpanzee etiquette at the London Zoo. Otherwise she would certainly not have survived the experiment, let alone succeeded so well that she became an international celebrity while still only a doctoral candidate. Today all zoologists, and many others, know the name of Jane van Lawick-Goodall.

On her first expedition[19] Jane Goodall, then unmarried, chose an entirely different method from that of Adriaan Kortlandt for observing chimpanzees. She was not working under time pressure and could therefore try to win the confidence of the animals to such an extent that she would be tolerated in their community and even treated as a friend.

To anticipate: the probation period she had to go through lasted a full eight months, and ultimately four years.

Whenever she approached a troop of chimpanzees, the animals hastily fled. In the beginning the apes fled to a distance of five hundred yards. After six months of daily confrontation they fled only when Jane approached within a hundred yards. After eight months the animals at last decided, apparently, to organize a trial reception. Here is Jane Goodall's account:

> It happened for the first time when I was following a group in the thick forest. The chimpanzees had stopped calling when they heard my approach, and I paused to listen, unsure of their whereabouts.
> A branch snapped in the undergrowth right beside me, and then

I saw a juvenile sitting silently in a tree almost overhead, with two females nearby. . . . Then I heard a low 'huh' from a tangle of lianas to my right, but I could see nothing.

Unarmed as she was, Jane Goodall sat down on the ground and bent her head, to avoid any appearance of aggressiveness. She continues her account:

For about 10 minutes these uneasy calls continued. Occasionally I made out a black hand clutching a liana, or a pair of eyes glaring from beneath black, beetling brows.

The calls grew louder, and all at once a tremendous bedlam broke out—loud, savage yells that raised the hair on the back of my neck· I saw six large males, and they became more and more excited, shaking branches and snapping off twigs. One climbed a small sapling right beside me, and all his hair standing on end, swayed the tree backward and forward until it seemed it must land on top of me. Then, quite suddenly, the display was over, and the males began to feed quietly beside the females and youngsters.

Jane Goodall had passed the test; she was no longer regarded as a menace and henceforth could stay in the vicinity of the apes.

To be precise, she was able to approach to within thirty yards of the group. Only after an acquaintanceship of four years was she able to move freely among them all.

Thereafter, when Jane van Lawick-Goodall[20] appeared on the scene, a few good friends ran toward her, bowed slightly, and extended a hand in greeting. As among men, she then had to place her hand in the chimpanzees' hands. Other members of the troop tolerated her, huffily ignoring her, and those who were in bad humor sulked.

This greeting by 'shaking hands' may strike the outsider as an incredible anthropomorphization. But that is not the case. Shaking hands plays a most important part in the community life of chimpanzees. If we observe all the occasions on which the animals make use of this gesture, we will obtain some idea of how the human gesture must have arisen. That is another story. It began with the discovery that savanna chimpanzees hunt other animals to eat their flesh. Hitherto science had considered anthropoid apes vegetarians. According to the latest researches, that is true only of the jungle chimpanzees, who can always find sufficient fruit and other tasty vegetable nourishment. Why, therefore, should chimpanzees in the jungle assume the risks involved in all kinds of hunting? Most animals—wiser than many men—follow the principle of avoiding needless dangers.

The situation is altogether different in the savannas and the plains. The eastern shore of Lake Tanganyika rises precipitously out of the water. Above a narrow strip of tropical rain forest tower treeless mountain slopes with numerous valleys and gorges. In these mountains plant food is scarce and must now and then be supplemented by animal protein. Here, and in the savanna-like transitional zone, are the chimpanzee hunting grounds. Boschboks, smaller antelopes, and river hogs are their game, sometimes other apes also, and occasionally a careless young baboon.

One day Jane Goodall observed four red colobus monkeys sitting on a tree. They were keeping an eye on an adolescent chimpanzee disporting in an adjacent tree. But since it did not approach threateningly, they thought themselves safe and neglected to look to the other side. Suddenly another adolescent chimpanzee appeared from that side. It ran with incredible speed along the branch on which the monkeys sat, leaped, seized one of the animals with both hands, and broke its neck.

As if they had been watching from a hiding place, five more chimpanzees immediately appeared, climbed the tree and advanced on the successful hunter, who unceremoniously tore his prey apart, and divided it among the others.

Such fairness is not always the rule. Grown male chimpanzees distribute their prey, if they do so at all, only as they please, according to their whims and preferences. Jane Goodall[21] once observed—using binoculars, for she still could not approach too closely—a vigorous old chimpanzee that had killed a boschbok, an antelope almost as big as himself. He sat contentedly in the grass enjoying his meal, his huge 'roast' tucked under his left arm. After each morsel of meat he ate a few leaves of 'salad' from a clump of twigs that he was holding in his other hand. When he was satiated, he wiped his greasy hands on a 'napkin' of large leaves.

There was good reason for his clasping his prey so tightly. For sitting all around him were several hungry members of the tribe eager for a share in the banquet. It was highly instructive to watch the principles that governed the distribution. Surprisingly, no one had a right to a portion, not even a chimpanzee higher in rank, before whom the successful hunter would ordinarily have to crouch humbly.[22] Once again it sounds like rash anthropomorphization, but we can scarcely avoid speaking here of first glimmerings of respect for the property of others. The higher-ranking chimpanzee, who flaunts his authority and power at a banana tree, loses his privileges before a hunter who has earned his prey.

Consequently, all the chimpanzees crouched greedily around the one who sat happily chewing and held out their hands, palm upward, in a typically human begging gesture—which henceforth, by all logic, can no longer be termed 'typically human'. Only a mother carrying a baby just a few weeks old received permission to eat when she pleased. All the others had to wait until the owner placed a morsel into their open hands.

An adolescent of about four received nothing. His begging became more and more importunate, until finally the owner of the meat slapped him for making a nuisance of himself. A cross old chimpanzee who must have been forty or fifty years old likewise went empty-handed. When he thought he was unobserved, he tried to provide for himself. He was caught, not by the owner himself, but by one of the group of beggars, who fell upon the thief and administered a good beating—presumably in order to ingratiate himself and be rewarded.

The whole scene suggests that shaking hands originated as a ritualized gesture of greeting: the one in need begs by holding out his palm. The other, if he wishes to, puts something into it.

The next phase is symbolic begging and present-giving. It may happen, for example, that two chimpanzees are approaching a papaya from different directions. In this case the higher-ranking animal has the right to take the fruit. But if the lower-ranking animal very much wants it, he runs up to the 'big shot' and before either of them has reached the fruit holds out his hand in the begging gesture. If the other is in a giving mood, he places his empty hand on the beggar's hand. This signifies that the fruit may be taken by the beggar.

There is even more to it than that. When Jane Goodall, who had meanwhile become an accepted member of the chimpanzee troop, was staying with the animals and wanted to play with an amusing chimpanzee baby, she had to follow a prescribed ritual. First she would extend her hand in the begging gesture in the direction of the baby's mother. A benevolent look from the mother served as the sign of her consent.

Two chimpanzees greet each other by touching hands. The one who extends his hand palm upward may be expressing submission or petition. (After H. van Lawick.)

Among two chimpanzees who meet one another, 'shaking' hands may also be understood as a kind of permission for the lower-ranking one to remain in the immediate vicinity of his superior. It is an act of friendship and hence a greeting in the fundamental sense of the word.

The Rain Dance

The beating of the thief, incidentally, was the most violent altercation that had hitherto been observed among chimpanzees living in the wild. Sometimes quarrels spring up from trivial causes but ordinarily the angered animals go no further than loud shouting and threatening gesticulations. Of the fifty animals whom Dr. Kortlandt observed in his six expeditions, none had bite wounds, injuries or scars. In general the interrelationships of anthropoid apes are so friendly that we human beings could well learn a lesson from them.

Anyone who witnesses an encounter between two groups of chimpanzees will expect the immediate outbreak of 'tribal warfare'. Instantly, both sides strike up an infernal din, as if they were trying to scare off a leopard. The males screech, drum resoundingly on hollow tree trunks, wave large branches in the air, jump up and down on two legs as if they were on a trampoline, and perform a kind of war-dance. But appearances are deceptive. For soon the racket subsides. Both parties approach closer. And then comes the astonishing denouement; two chimpanzees suddenly hold out their hands, two others actually embrace or exchange a kiss. It looks as if they are genuinely overjoyed by a reunion after a long separation.

The fact is that all the chimpanzees in a sizable area know one another personally, though they belong to various groups. Each animal adheres to another group from time to time.

After meeting, both groups stay together for several hours, sometimes for days at a time. When the groups part again, the membership of each is almost always somewhat changed. Perhaps one chimp has been offended too often and hopes for a pleasanter life in the other group. Or else two members of different groups have formed a close friendship during the meeting and decide to continue on their way together. Occasionally some of the animals will leave the group entirely and live a solitary life for a time.

Precisely regulated rankings exist among chimpanzees. But no individual has the right to interfere in another's personal affairs. Chimpanzees are herd animals to a far smaller degree than man, says Dr. Kortlandt. They love to stay together, but each one follows his own bent. It is truly astonishing to discover such strongly marked individualism among anthropoid apes.

What, then, is the purpose of the martial demonstration when two groups meet? Certainly an element of fear is present, coupled with aggressiveness. Perhaps we may compare it with the feelings of soccer players before they begin a game with another team. In the main, however, the whole display is no more than an attempt to impress, possibly in order to clarify the rankings of the individual members of each group, which may have been forgotten by the members of the other group.

On the whole, rank among chimpanzees is not established by fighting, not even by a more or less playful and sportive wrestling match, but by pure showing off, sheer displays of strength, usually ostentatious bending of tree trunks. This kind of behavior starts early in life. Adolescents delight in such trials of strength. They confront each other with chests puffed out and rattle a few thin branches—conduct that always struck Dr. Kortlandt as rather ludicrous. Stronger chimpanzees simply ignore such puny displays. But when the Tarzan of the group performs his impressive feats of strength on the trunk of a tree, all the rest stand around as if at a sporting event, watching in amazement from a respectful distance.

From time to time every chimpanzee has his fit of rage. He must discharge his accumulated aggression. Unusually hot-tempered human beings have a similar impulse, as is well known. At certain intervals they spontaneously have outbreaks of rage inexplicable on rational grounds. Then they stand in the same posture as a chimpanzee trying to impress his fellows by threats. They bellow senselessly at subordinates, harry neighbors, whip their children, or vent their anger upon their innocent wives.

Chimpanzees behave much more rationally in this situation. As Dr. Kortlandt observed in the Congo jungle, the particularly muscular male apes are the ones likely to have fits of fury in placid times, without any visible cause. The other members of the band notice at once when this mood is on him and carefully keep out of his way. The chimpanzee in a rage has enough sense not to go chasing after a 'scapegoat'. Instead he contents himself with a tree as surrogate, just as he does in performing his feat of strength. He rocks the tree back and forth, uproots it and tears it apart. By then he has calmed down, and goes peaceably about his business.

This process of gaining prestige by histrionic displays and knocking about inanimate objects can sometimes lead to curious shifts in power. In the band with which Jane Goodall had struck up a friendship, the chimpanzee she called Mike was 'lower middle-class'. One day he came upon a heap of empty gasoline cans at the scientist's encampment. By chance he had a sudden fit of fury, and for want of a tree hurled the cans around and dragged some of them over the ground, making a tremendous racket. The spectacle so enormously impressed the other chimpanzees that they henceforth, without more ado, acknowledged Mike their 'Number One'.

Of course chimpanzees also fly into quick rages when they do not like something—for instance, the weather. During the rainy season, when it seems as if it will never stop pouring, the male apes' initial apathy can change within a few minutes to a mixture of impotent fury and community display of strength—a ritual that almost suggests a deliberate conjuring of the spirits of nature. Jane Goodall witnessed this strange spectacle four times.[19] She calls it the rain dance.

As she describes it, rain had been falling all morning long. Sulkily, the sixteen chimpanzees in the band descended from the trees and crouched side by side on a steep mountain slope. They held their bodies bent far forward over their knees, their heads dangling down.

But toward noon the rain poured down and the thunder cracked more violently than ever. Abruptly, a muscular male chimpanzee sprang to his feet screaming, drummed on the ground and struck out at branches. In a moment his rage had infected six other males. They formed into two groups and one group after the other stormed up the steep grassy slope between the trees.

Close to the ridge of the hill, one male suddenly turned around and rushed at top speed diagonally down along the slope, screeching loudly and beating a branch against the trunks of the trees. Then the first member of the second group raced down in the same fashion, crossing his predecessor's course. Meanwhile the females and young climbed into the surrounding trees and looked on.

The next performer was already standing on the ridge, swaying back and forth with exaggerated movements, and swinging his arms. Then he too came speeding downward. The others had climbed trees, and now leaped to the ground from heights of twenty feet and more, tearing off branches as they fell and dragging these behind them during their screeching descent of the slope.

At the bottom, each male retired into a tree to catch his breath for a moment. Then the game began again from the beginning, while the rain beat down more and more, while lightning flashes came almost without pause and the thunder drowned out the screams of the chimpanzees.

Suddenly, after about fifteen minutes, the whole show stopped as abruptly as it had begun. The spectators descended from the trees, drifted toward the top of the hill, and vanished down the opposite slope.

Impotent rage against the powers of nature expressed in common action—such is the basis of the rain-dance ritual. From this to the dances of savages for the purpose of conjuring spirits is only a small step.

Studies of the chemical components of the blood have shown[23] that the chimpanzee is much more closely related to man than to any other species of monkey or anthropoid ape. The behavioral investigation cited above reveals more aspects of this kinship.

What had we known previously about our closest relatives in the animal kingdom? Nothing more than that they made faces behind the bars of zoos, that in confinement they reacted foolishly or cholerically, or showed signs of sexual degeneracy. Consequently, people were shocked by Darwin's theory of the descent of man. It is incomprehensible that we have had to wait for the sixties of the present century before a few scientists began observing the chimpanzee's manner of life under free, undistorted, natural conditions. Their scientific conclusions have in a sense redeemed the honor of our ancestors. After all, Jane Goodall was treated far better in chimpanzee society than any chimpanzee has ever been treated in human society!

But however close or distant the kinship between man and chimpanzee may be, however many similarities in behavior they may show, one thing is evident: these matters provide no clue to other phenomena, such as language and intelligence, or other patterns of social conduct, such as aggression and, above all, the 'predator components' which are decidedly stronger in man than in the chimpanzee. The behavioral psychologist must turn to altogether different species of animals, species seemingly much more remote from man, to investigate other specifically human traits. What he finds is sometimes astounding.

1 *Swinging a club and assuming an erect posture, a chimpanzee dashes past Dr. Kortlandt's and Dr. Van Zon's hiding place to attack a stuffed leopard (not visible in the picture).*

2 *An adult chimpanzee is twice as strong as a man and has teeth scarcely inferior to a leopard's.*

3 *At the edge of the jungle the chimpanzee troop peers out to see whether the banana plantation is free of enemies. At far left a forty-year-old senior citizen crouches—no longer the strongest, but still leader of the band on account of his experience.*

4 Four bottlenose dolphins performing an aerial leap with the precision of trapeze artists. Each member of the team can rely on the other—to the fraction of a second and of an inch.

5 (ABOVE) *Like a faithful dog, the dolphin brings its mistress a basket.*

6 (BELOW) *The dolphin is being trained to deliver supplies to underwater stations. In his leisure time he plays headball with a two-hundred-pound turtle.*

7 *The stork scarcely has the voice for love songs. In moments of ecstasy he settles on his bride's back to bill if not to coo.*

8 (ABOVE) *Singing lesson in the long-tailed titmouse's kindergarten. The fledglings are imitating what the father (outside the picture) has just chirped to them. Young bullfinches (BELOW) (9) will be faithful their whole lives long to the melodies they have learned from their father. They will pass these on to their offspring.*

10 *Scientists come to spend nine months in the icy Antarctic night studying emperor penguins.*

11 (ABOVE) *The first squadron of emperor penguins has landed on an ice floe.*
Some days later their number will have swelled to some six thousand.

12 (BELOW) *Two souls with but a single thought. When a baby penguin*
cries, both parents always do the same thing simultaneously.

2. The Road to Speech

The Intellectuals of the Sea

In the vicinity of the Lesser Antilles a young dolphin had wandered far out of sight of his band when he was suddenly attacked by three sharks. At once he uttered a series of shrill whistles: SOS signals in dolphin language. The short twin whistles sound like an overwound alarm siren: the first part rises sharply in pitch, the second half falls just as abruptly.

The effect was extraordinary. The twenty-odd dolphins in the school, who were conducting a lively palaver with whistling, squeaking, grunting, gurgling, booming, and peeping sounds, immediately stopped 'conversing'. As is the case when distress signals from a ship at sea are heard, absolute 'radio silence' prevailed. Then the animals shot toward the scene of the attack at their maximum speed of nearly forty miles an hour. Without slowing their speed, the male dolphins rammed the sharks. Again and again they crashed into their sides, until the sharks, their cartilaginous skeletons shattered, sank lifeless to the bottom of the Caribbean Sea.

During the fight the females went to the aid of the severely wounded young dolphin, which could no longer rise to the surface by its own strength. Two females ranged themselves on either side of him, thrust their flippers under him, and raised him so that his blowhole was above water again and he could breathe. This rescue maneuver was carefully regulated by an exchange of whistling signals. From time to time the 'stretcher-bearers' relieved one another. On one occasion scientists have observed such aid being continued without a break, day and night for two full weeks, until the injured dolphin had recovered.

From these and other observations by the American neurologist Dr. John C. Lilly[1, 2] the dolphins have begun to appear, since the beginning of the sixties, as modern equivalents of the ancient fabulous beasts. Half-jokingly some people[3, 4] have suggested that these 'prodigies of the seas' may be superior to man in linguistic virtuosity and intelligence. But in all seriousness, the psychiatrist Professor G. Pilleri[5] writes that the dolphin's brain 'attains a degree of centralization far beyond that of man'. In the opinion of this scientist 'the ultimate status of man's brain in the ranking of mammals is today beginning to be a matter for doubt'.

If man should ever learn how to talk to dolphins, the physicist and biologist Leo Szilard[3] has predicted, these 'intellectuals of the sea' would win all the Nobel Prizes for physics, chemistry and medicine, and the Peace Prize to boot. In point of fact, what the 'delphinologists' have actually discovered so far are only a few grains of truth which, while they do not exclude such fantastic prospects, still are far from confirming them.

Nevertheless, when two dolphins 'talk' to one another they engage in an actual dialogue. Dr. Lilly in his Institute for Communications Research on the island of St. Thomas in the Antilles was able to provide fairly convincing evidence of this when he divided the dolphin pool by a panel of sheet metal, thus separating a dolphin couple. First there was a shrill concert of whistles by both animals. They could recognize each other's voices, but could not see each other. They tried to leap up high so that they could see behind the partition, but in vain.

Then both sank into gloomy silence. By and by the male began encouraging his mate to conversation again. Long monologues were necessary before the female once more made a sound. At this the male fell silent, and 'spoke' again only after his mate had finished. This went on in constant alternations of sound production for periods of up to half an hour.

Sometimes there would be a duet, when one dolphin chimed in with the other's whistles, now loud, then again quietly, sometimes reaching high into the realm of ultrasound, at other times in a low, grunting bass.

Dr. Kenneth S. Norris at the Makapuu Oceanographic Institute in Hawaii made use of these animals' love for dialogue to have his Pacific dolphins talk by telephone with Atlantic members of the species in the marine laboratories of Miami. Communication via underwater microphone, public telephone cables, and an underwater loudspeaker for each of the participants yielded far better results than expected. In this case, too, each dolphin let the other finish what he had to say before beginning to reply with his gurgling and whistling sounds. Evidently the dolphins of all the oceans speak the same language.

The loud whimpering and yowling of dolphins during mating season is reminiscent of the moonlight sonatas performed by tom-cats on a roof. Professor Winthrop N. Kellogg[6] of Florida State University has recorded the dolphin sounds with the underwater listening apparatus aboard his motorship. The most curious factor is that a couple can chat even if each mate is separated by a considerable distance in the midst of a talkative school. Each dolphin always knows who is addressing whom. The one directly addressed replies in his turn and is not in the least bothered by the general conversation among the others. The scientists have dubbed this the 'cocktail party effect'.

The suspicion that these reciprocal utterances might well be a kind of language was reinforced when Dr. John Dreher, Dr. William E. Evans, and Dr. J. H. Prescott[7] of the Lockheed Aircraft Corporation listened in on five bottlenose dolphins. They placed fifteen buoys to make an obstacle across the mouth of Scammon Lagoon, which is on the Pacific about three hundred miles south of San Diego, California. Late one afternoon the five dolphins who made their home in the lagoon, and who were just returning from a protracted expedition into the ocean, sighted the buoys. They paused abruptly, turned away, and gathered in the safe and shallow waters of the shore.

Soon a scout detached himself from the group and swam cautiously from one buoy to the next. When he returned to his companions, there was an excited burst of shrill whistles. The result of this 'discussion' was that a second dolphin swam off to inspect the disturbing objects. On his return there was another vehement session of whistling. Only then were the dolphins reassured. Cautiously and silently, they moved forward past the buoys and vanished in the lagoon.

Of course the scientists had no way of knowing in detail what the animals told one another. Two other scientists had the good fortune to gain some deeper insights toward the end of 1965. They are the engineers T. G. Lang and H. A. P. Smith[8] of the U.S. Naval Ordnance Test Station in Pasadena, California. The American Navy, it seems, is not quite comfortable at the thought that there may be, along with its nuclear submarines, creatures cruising the seas who possess better methods of underwater navigation and communication than the Navy itself. It is therefore devoting considerable efforts to investigation of dolphins.

The telephone conversation between the dolphins in Hawaii and Florida inspired the two naval engineers to try the following variation. They placed two large bottlenose dolphins, Doris and Dash, who had recently been captured in the Pacific, in two separate soundproof tanks and provided them with a private underwater telephone. The experimenters were able to interrupt the telephone connection whenever they pleased. Every sound made by the two animals was recorded separately on different tracks of the same tape.

The female Doris and the male Dash were instantly aware of when the connection was functioning and when it was not. In their constantly alternating exchange of sounds, the dolphins expressed themselves very tersely. Neither partner permitted himself to go on talking for more than four or five seconds. If there was no reply, he dropped into silence. Each animal alone would utter a few sounds at sizable intervals— probably only to check whether the other was again 'on the phone'. On the whole the lady dolphin proved to be by far the more talkative of the pair.

The following sounds were noted: clicks, grunts, and squeaks. Probably these are signs of emotional states, expressing anger, irritation, and comfort. Evidently they do not serve to exchange information, for the language of the dolphins is a whistle language.

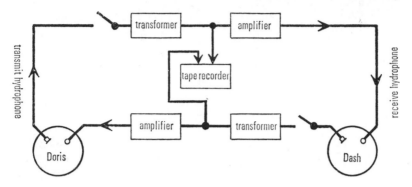

The telephone connection between the tanks of the bottlenose dolphins, Doris and Dash.

The whistles fall into six different types. The two Americans offer the following hypotheses in regard to their meanings:

Whistles in Group A have a simple sound pattern, are always the same, and are used by both animals in the same way. These whistles have been interpreted as the questing call to initiate contact: 'Hello? Is anybody there?' Whistles of Groups B and D are somewhat like the SOS cry described earlier; they rise sharply in pitch, then drop off just as abruptly. There are two versions of the same basic signal. Doris produced only the higher-pitched B whistles, Dash only the lower-pitched D whistles. Possibly this is a matter of individual variation serving as a sending signal to identify the particular dolphin speaking. Every conversation is peppered with these 'identifications' sent out in rapid succession. This would explain why many animals in a school can talk simultaneously and yet each follow the voice of its partner in the medley. That is, B and D whistles can be interpreted as: 'Doris speaking' and 'Dash speaking'.

Whistles of the types C, E, and F seem the most interesting. They are exceedingly rich in variations of tonal color and intensity. Moreover, in the Lang–Smith experiment they were uttered only when the telephone was functioning and the dolphins had already established contact by search and identification calls. The F whistles in particular

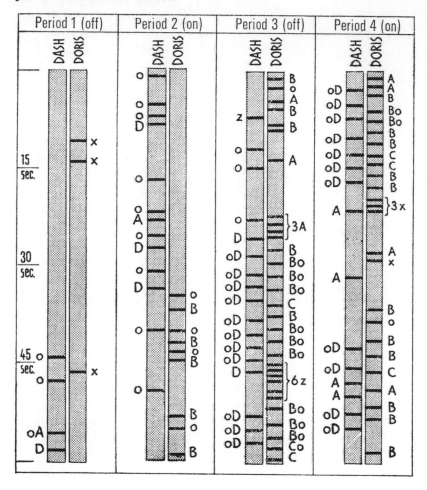

The first four minutes of the telephone conversation between the
bottlenose dolphins Dash and Doris. Detailed explanation in text.

A = *Hello? is anybody there?*   F = *Meaning unknown*

B = *Doris speaking*   o = *Echo location signal*

C = *Meaning unknown*   x = *Click*

D = *Dash speaking*   z = *Squeak and grunt*

would be eminently suited for the exchange of information. These whistles rise in pitch, then drop off, and end with a 'coda' of ever-varied tones. Usually they conclude an extended series of 'remarks'.

A dolphin conversation, according to Lang and Smith, takes the following course (see Figure on p. 30): in alternate minutes of the tape recording the telephone connection is interrupted. During this period the 'Hello? Is anybody there?' signal sounds twice. Four times the male calls out at relatively long intervals: 'Dash speaking,' the female five times: 'Doris speaking.' That is all.

Then the connection is established. Dash suddenly hears: 'Doris speaking. Hello? Anybody there?' He answers first with a squeak of pleasure. Immediately afterward his 'sonic locator' ticks; he is trying to locate his partner.[9] Doris hears this and hastily calls out four times in succession: 'Hello? Anybody there?' She receives the answer: 'Dash speaking.' Simultaneously she also ticks her sonic locator and repeatedly gives her personal identification signal. Within seventeen seconds Dash likewise identifies himself no less than nine times in succession. But Doris cannot locate him, for he is swimming in the other tank. She reacts with six angry grunts.

Desperately, Dash leaps into the air several times in order to peep beyond the tank wall. But within four minutes he realizes that this action has no hope of success. After eight minutes he stops his echo soundings, apparently recognizing them as pointless.

Let us imagine two human beings who have never heard of telephones, microphones, or loudspeakers placed in the same situation—say, two Papuans from New Guinea. How long would they run round and round the telephone booth before they realized that they were dealing only with a 'ghost voice'? The dolphins, however, adjusted to the new situation in seventeen minutes and a few seconds, whereupon they commenced a 'rational' conversation with F whistles.

A further experiment that took place four months later proved that the conversation was indeed 'rational'. The two engineers decided to fool Dash, and played to him the tape of his conversation with Doris. At first Dash answered every sound on the record willingly, and much as he had done the first time. But suddenly, after seventeen minutes and fifty-three seconds, he abruptly fell silent and refused to utter another sound. Next day the same experiment was repeated—and once more the dolphin went on strike at very nearly the same point in the text.

How had he discovered the deception? Immediately before Dash broke off his dialogue with the tape, there had come for the first time a series of seven F whistles rich in variations. The scientists had previously assumed that these whistles served to communicate information. And it was precisely at this point that the dolphin became wary.

That fact signifies something altogether unique in the animal world. Sounds of this sort evidently have meaning only in connection with remarks that the interlocutor has made immediately before. We are tempted to say that the dolphin put an end to the conversation because he suddenly realized that the talking tape was being silly.

Encouraging though these experiments were, at the present time all attempts to decipher the content of the C, E and F whistles—in other words, to plumb the complexities of dolphin language—appear to be hopeless. Man can certainly fathom the meaning of animal signal cries of the kind that prompt members of the species to flee, coax them to approach, anger them or placate them, or induce some other prompt and clearly recognizable reaction. But when a dolphin merely takes note of what he has heard and acts on it later, or when he 'only' thinks and doesn't do anything at all—how can a human being ever arrive at the meaning? The more like human speech an animal's language is, the harder it will be to decipher.

To this day the language of the Etruscans is a sealed book to us. How much more complicated will it be to 'translate' dolphin language. In 1963 two electronic engineers, Leo Balandis and George Rand of the Sperry Gyroscope Company, programmed a computer to help them decipher the mysteries of dolphin whistles. But no electronic brain has yet accomplished miracles, and five years later no significant results have been reported.

For this reason several scientists are trying to initiate a conversation with dolphins by another technique. Since there seems no way for them to learn 'dolphish', they wanted to teach the dolphins human language. Such a program can succeed only if dolphins are actually more intelligent than men.

Dr. John C. Lilly[10] had noticed that his dolphins sometimes imitated whole sentences he had spoken, and to some extent actually used them in meaningful contexts. Their pronunciation sounded rather like Donald Duck's, but they had a vocabulary far exceeding that of a gifted parrot. When the dolphins were in good humor, they would instantly imitate virtually anything Lilly said—although their speech sounded like a tape running at excessively high speed. Evidently dolphins have an extraordinarily fast speech tempo. They have obvious difficulty in producing sounds as slowly as human beings produce them.

'One conclusion that may be drawn is that it may be easier to teach a dolphin a human whistled language than the more common human vocalization,' Dr. R. G. Busnel[11] proposes. There are three regions on earth where men communicate and even converse by whistles over long distances. The peasants of a village in the French Pyrenees talk from pasture to pasture, across an intervening valley, in a whistle language called Aas. On one of the Canary Islands the inhabitants put four fingers into their mouths; their loud whistle language is called Silbo Gomero. In the western Sierra Madre of Mexico the whistle language is known as Mazateco. One of these human whistle languages could certainly serve as the basis for another attempt at communication with dolphins.

In the meantime Dr. Dwight W. Batteau[12] in Hawaii has carried out a similar idea. He has developed an electronic apparatus which converts all sounds of the Hawaiian language into equivalent whistles. His two dolphins, Puka and Maui, already understand simple commands from the whistle apparatus, such as: 'Maui, jump through the hoop!' or, 'Puka, repeat the word hoop!' They promptly follow these instructions. Apparently the animals have no difficulty in whistling sounds like those of the whistle apparatus. The scientists believe they can recognize certain 'words' used by the dolphins as imitations of the electronic vocabulary. But this has to be proved scientifically. Therefore Dr. Batteau is now constructing an apparatus that will convert the sounds of the whistling machine and the dolphin imitative whistles back into human language. Perhaps this apparatus will clarify the question of whether dolphin or man has the greater linguistic talent.

The question is, why should dolphins have a language similar to man's in its complexity, since all other animals manage without complicated verbal communication? The answer may possibly lie in the hunting tactics and social conduct of these marine mammals. Schools of dolphins launch well-organized 'battue' hunts of schools of fish. They can drive their prey together like cattle, encircling them on the surface of the water in the open sea, or forcing them into inlets where they can gorge on them. They can slit open filled fishermen's nets and steal the contents. But they can also collaborate with human beings and drive schools of fish into the nets. In all such tasks it is obviously necessary for the dolphins to work in concord, so that each one can depend on the others. They seem to manage this by sonic signals.

It is true that lions (see page 126) and wolves also work together in a tactical plan to encircle their prey, without having to use language—aside from a few 'identification' and 'ready for attack' signals. Land predators in their maneuvers over considerable distances largely guide themselves with their eyes or their noses. In turbulent water, that is wholly impossible. We can conclude, therefore, that the hunting technique of dolphins virtually depends on transmission of complicated messages from animal to animal by means of language.

On October 18, 1967, the Associated Press reported from Moscow that a school of dolphins in the Black Sea 'asked for help' from a fishing boat. Near the Crimean coast the small boat was surrounded by several dolphins and pushed in the direction of a buoy. There the Russian fishermen found a young dolphin that had been caught in the anchor rope of the buoy. The men succeeded in freeing the dolphin baby. With whistles of joy the members of the band greeted their uninjured offspring, and escorted the fishing boat all the way into the port.

The 'intellectuals of the sea' would scarcely have ventured a similar thing off the Swedish Baltic coast. For there the fishermen wage war—though not a very effective one—against the dolphins because of the damage they do to the fishermen's nets. As soon as a school of dolphins crosses the path of a fishing vessel, a harpoon boat is summoned by radio. The harpooners, however, can make only one attack. Thereafter the dolphins are able to distinguish the harpoon boat from the fishing boats, although both are the same type of vessel.

Dolphins are thus quite able to learn from experience. There are, moreover, many indications that they can transmit their experiences to other dolphins—perhaps by means of their whistle language. How otherwise can we interpret these two contrasting events in the Baltic and the Black Sea?

Language, moreover, seems to release forces in dolphins that are more powerful than the most elemental instincts. When one of the largest and most dangerous predators in the world, a thirty-foot-long killer whale, dashes in among a school of dolphins, who break into shrill whistles of alarm, it might be thought that the beasts would all flee in panic—as men often do in a similar situation. But that is not what occurs. First the dolphins try to rescue injured companions; only then do they make off.

In a great many respects the behavior of dolphins presents us with enigmas. For example, there is their by now legendary friendliness toward human beings, attested by numerous instances of dolphins allowing men to ride them through the waves for sport, or rescuing drowning men. There is not the slightest constructive reason for this hereditary friendship; on the contrary, men have all too consistently persecuted, captured, shot, and killed dolphins. Nevertheless, not a single case has ever been recorded of a dolphin's making a hostile gesture toward a man—not even when the man is engaged in killing the dolphin. In dealings with our species the animal seems to lose all instincts of self-preservation, self-defense, and reprisal. Yet it has those instincts; as we have seen, it is quite capable of harassing sharks to death.

But it is not for insight into these questions that large sums are being poured into dolphin research in the United States and the Soviet Union. On the contrary, the aim of these studies is to harness another 'super-human' ability of these animals for military purposes. That is their ability to 'hear pictures' by means of ultrasonic waves.

In spite of enormous efforts by electroacoustic engineers, no one has yet succeeded in constructing an underwater sonic apparatus that functions even approximately as well as the built-in apparatus the dolphins possess. The animals cannot only detect the presence of something at great distances; they can distinguish with their ears between a herring and a piece of driftwood, between a perch and a barracuda, and they can locate a blue whale just as well as a ball of shot thrown into the water. Moreover, no humanly constructed jamming apparatus can interfere with the dolphin's sonar.

In other words, a dolphin would be an ideal 'auxiliary apparatus' for nuclear submarines. In 1965 Dr. Kenneth S. Norris[13] successfully trained a dolphin to follow a motorboat out into the open sea, to respond to commands by making precise changes of course, and to bring supplies to the crew of an underwater station at depths of two hundred feet. Practically speaking, with training such as this, it would be feasible for 'man's first sea-dwelling domestic animal' to be admitted or released from a submarine through some sort of torpedo tube, to learn to locate and report the dolphins of enemy vessels, to learn to identify hiding places at the sea bottom, and to be used as pilots.

More Musical than Man

If crested larks could test the musical aptitude of human beings with scientific accuracy, they would probably give Toscanini a high mark, but most people would rate far beneath the lark level.

This remark is not some clever quip of George Bernard Shaw, but the sober result of sonic-spectographic studies of two crested larks living in the fields on the outskirts of the city of Erlangen in Germany who resoundingly imitated whistles from a shepherd.[14]

It may be startling to learn that crested larks (not meadowlarks) have the ability, like parrots, ravens and mynas, to imitate human voices. But zoologists have recorded one crested lark[15] that could sing seven human songs and could pronounce a few words, as well as the numerals 'one, two, three,' in a rather thin voice, but quite intelligibly.

Lacking any closer contact with human beings, the two Erlangen larks could only sharpen their skill on the whistles a shepherd used for giving orders to his dogs. These whistles had been carefully codified. A rising 'scale' of five whistles, the last of which suddenly dipped far down in pitch, meant 'run away!' One, two, or more sharp whistles signified different degrees of 'hurry, hurry!' If a long-drawn-out whistle quiveringly changed pitch, the dogs stopped whatever they were doing. Three repetitions of this signal meant, 'Come here!'

The larks had mastered all these signals. It must have been most confusing for the dogs to be whistled back and forth by the birds and their master. The dogs therefore formed the habit of throwing a quick look at their master after each whistle, and carrying out the order only if he confirmed it by moving his hand or nodding his head. When the dogs were treated to a tape of the lark whistles, they behaved in the same way.

Even more astonishing facts were revealed when Dr. Erwin Tretzel, the Erlangen ornithologist, made a careful sonic-spectrographic comparison of the shepherd's whistles with the larks' imitations. For it turned out that the shepherd was horribly unmusical. He hardly ever hit the same pitch twice, and he certainly had no feeling for proper time. The little birds had no such difficulties. They reproduced the tunes they had learned with unvarying exactitude, using the whole-tone intervals of the C-major scale with never a flat note.

These findings led to a further question: If there are always such flaws in tune and rhythm in their model, how do the larks establish a standard for their own performance? It may seem fantastic to say, but the birds clearly act on their own initiative. They transpose what they have heard in a manner that suits their own dispositions. As Dr. Tretzel explains it: 'The lark had grasped the "idea," the ideal form of this motif, and whistled it as the shepherd probably thought he was rendering it but seldom actually succeeded in doing.... The lark produced all the shepherd's whistles far more purely and musically, more delicately in tone and more elegantly in scale. In musical terms, it refined the whistles, as it were. Here an astonishing feeling for form and metrics has been displayed by a bird that no one has hitherto thought a good singer. Certainly no one would have suspected that there were principles of order underlying the jabbering medleys it produces.'

We may assume that these crested larks of Erlangen are not especially gifted birds. They are by no means individual cases, prodigies whose accomplishments are cited to prove the musical superiority of birds. In 1966 Dr. Erwin Tretzel[16] observed the same phenomenon in perfectly ordinary blackbirds in home gardens of Garmisch-Partenkirchen. There, some of the blackbirds imitate the whistles with which a house-holder calls his cat. This was a highly dangerous hobby for the black-birds, for a cat was observed leaping at a bird which was mimicking such a call. But the human melody evidently had an irresistible attrac-tion for the birds. However, they transposed the theme up a fifth, where apparently it suited them better. Aside from this change, 'the blackbirds were far superior to the human model in constancy of rhythm and frequency'.

Human ears seldom notice this virtuoso musical talent. It is especially hard to discern in birds that twitter very rapidly. The wren,[17] for example, sings no less than 130 notes in seven seconds! We can grasp such a wealth of notes only with electronic instruments.

Although our human senses cannot deal with it, these series of rapid notes hold definite meanings for birds. This was discovered by Pro-fessor W. H. Thorpe of Cambridge.[18] In the dense rain forest of Uganda live pairs of black-headed gonoleks (Lanarius erythrogaster). In order not to lose each other in the impenetrable network of branches, the male and female continually sing a duet. At first the English ornithol-ogist thought the song was coming from a single bird. But when he happened to step between the two singers, he realized that the 'yoicks' in the first half of the song came from in front of him and the hissing conclusion from behind.

The two parts of the song flowed into one another without a break— at least to our ears. But the birds recognize a tiny pause, and oddly enough, this pause is what counts. In order to avoid confounding their mates with neighbors, each pair has its own pause length as an identification signal. For one pair Professor Thorpe measured an inter-val of 144 milliseconds, whereas the neighbors' pause lasted 425 milli-seconds. The latter pair maintained their pause with a constant accuracy of 4·9 thousandths of a second.

Thus the acoustic reaction and reception time for these birds is three times as rapid as that of man.

How superb, then, must be the feeling for melody and rhythm in birds whom we universally recognize as good songsters! Johannes Kneutgen[19] at the Max Planck Institute in Seewiesen, undertook the following experiment with the South-east Asian dayal or magpie robin, one of the best of avian singers. While the bird was chirping, he placed a ticking metronome close beside it. The dayal reacted to the ticking like an opera singer to her conductor, fitting the tempo of its song precisely to the beat of the instrument.

When Kneutgen slowly but steadily increased the tempo, the bird tried to keep pace. Finally, however, the dayal apparently came to the 'conclusion' that the motif it was singing should not be speeded up any further. Abruptly, it shifted to another of the melodies in its repertory, one whose natural tempo corresponded to the speed of the obtrusive instrument. Could there be any better evidence that birds have a feeling for the beauty of musical forms?

Many specialists already agree on one point: that the songs of birds do not have merely utilitarian ends. They do not serve only as war chants to frighten off rivals and delimit territory, or as love songs to lure the female. There is far more than that behind them.

When a nightingale breaks into soaring coloraturas of an aesthetic quality far beyond the needs of mere communication, when a crested lark musically refines a shepherd's whistling, when a blackbird in a mood of contentment, desiring nothing and having nothing to fear, plays artistically with tones, when the American wood pewee strikes up a song which resembles the first main theme in Beethoven's Violin Concerto—are these birds producing something that can be regarded as a preliminary step toward art? After decades of investigation many scientists believe that such an assumption is justified.

Professor Konrad Lorenz[20] comments on this point:

We know that birdsong reaches its highest perfection and its highest virtuosity where it does not serve the functions of delimiting territory, luring the female and intimidating a rival. A bluethroat, a dayal, a blackbird, sing their most artful songs only when they are singing creatively to themselves, in an altogether temperate mood. When the song is aimed at some purpose, when the bird is singing at an opponent or making a display for a female, all the higher subtleties are lost.

The formally most perfect, tonally richest form of song is then mangled and inhibited by the rising sexual instinct, and only a few loud motifs are used for the territorial song which serves sexual and sociological needs. The singing bird achieves its highest artistic accomplishments in the very same biological situation and the very same mood as man—when he is in a state of psychic balance, remote from the seriousness of life, as it were, and creating in a purely playful manner.

It might sound like a bad joke to assert that birds hold anything like 'song recitals'. To be sure, there is considerable evidence that the avian artists feel their own compositions to be beautiful. And the female apparently shares this opinion. But for competing neighbors the song has definitely aggressive implications and serves as a deterrent. Nevertheless, ornithologists have been able to prove conclusively that 'singing for an audience' does take place among birds.

Several species of strawberry finches, which are native to southern Asia and Australia, offer such remarkable performances. These brilliantly colored birds form large flocks and brood in colonies of sometimes several thousand in one tree. In contrast to the songbirds of Europe, they have no territory to defend. Their problem is rather to get along well with one another. Any song with warlike connotations would destroy their community life. By a magnificent shift in meaning, these birds have therefore developed singing into a friendly ritual.

Before nightfall, one or the other of them always gives a solo performance. The neighbors silently come close and listen. In this way the strawberry finches meet one another and become friendly. Here is an ideal example of the unifying power of beauty!

It is time men freed themselves of the notion that a feeling for the beauty of musical and rhythmic forms is inseparably and exclusively associated with the nature of humanity. Why should not birds be better than we in this realm? Singing and musicality are not part and parcel of human survival. We exchange information through the use of fairly monotonous sounds, not through musical elements, as birds do. If only on these grounds it is fitting and natural to concede pre-eminence to the feathered singers.

Birds are, however, no peer to men in musical memory. A well-trained bullfinch can manage several short songs. Hagenbeck's famous myna speaks twenty-two sentences. More than that simply will not fit into birds' small brains. No feathered vocalist would be capable of acquiring Schubert's song cycles, let alone the repertory of a popular singing star. But that is, as we have said, a matter of memory, not musical talent.

Birds Call One Another by Name

The bird's superior musicality, or what we admire as such, is viewed by many scientists, such as the Viennese linguist Professor Friedrich Kainz,[21] as a defect. These scientists maintain that all the bird is capable of producing is a kind of mood music, an interjection, nothing more.

When, for example, a chicken notices a hawk circling in the sky, it is frightened. Quite unconsciously and unintentionally, a squawk of horror escapes it. This 'music' frightens all the other chickens that hear it, so that without having seen the hawk themselves they will run for cover. Such is the function of cries of alarm for many animals. It is thus of a different nature from the utterance of human beings when they give an alarm.

The same sort of expression of moods by 'music' may possibly be involved in the coaxing sounds that indicate food, in the cries of fury directed at predatory birds, in cries of forlornness, in threats of attack, in attempts to placate enemies, and in manifestations of similar emotions. But in behavioral research it is well to be cautious about generalizations.

If a female daw[22] wishes to fetch her enterprising mate home from the fields late in the evening, she flies low over him and from behind calls out a piteous 'kyoo'. That is intended to make him anxious and induce him to fly home with her. In this respect, then, the call is 'mood music'. But in this case the daw undoubtedly sounds it deliberately and directs it at a specific individual—so that we have here an exceedingly important step in the evolution of speech!

Critics like Professor Kainz are actually making quite another point: that in contrast to man, birds have no words, no sentence structure, no syntax. This is largely true. But not entirely. For investigators of animal speech have come upon some remarkable facts.

A deserted male bearded titmouse usually announces his desperate plight to the world at large. He indefatigably chirps, 'Chin—jick—chray'. This song is one of the few bird melodies whose 'verbal' meaning we have been able to decipher, because it is compounded of individual calls with which we are already familiar.

As the Vienna behavioral scientist Professor Otto Koenig[23] has established, 'chin' in titmouse language means approximately, 'Attention, alarm!' 'Jick' expresses a lively desire to mate. 'Chr' is used to lure females, and 'ay' is the cry of desertion and plea for pity. Thus the song passage might be translated: 'Attention, I'm in a mood for love. Woman, come here. I'm so lonesome.' Titmice already living together as established pairs call out merely: 'Jick—chr'—that is: 'I'm in a mood for love. Mate, come here.'

Ants, too, can probably combine single 'words' in various contexts to produce different statements. Ants communicate by means of odors. They possess some half a dozen scent glands, each of which produces a 'fundamental concept' such as 'Alarm! Enemies entering nest!' or 'This track leads to the source of food'.

Professor Edward O. Wilson,[24] the American entomologist, has found indications that ants can combine several odors to make mixtures. In this way they possess more 'words' in their vocabulary than the mere number of scent glands in their bodies. Apparently these insects can also emit their signal odors at different rates of speed and modulate each emission so that odors of varying strengths result. Thus they create a kind of Morse code. Under these circumstances the existence of a form of sentence structure, though one very alien to our notions, is at least conceivable.

In bee language, too, one utterance is by no means rigidly confined to a single situation. As Professor Martin Lindauer[25] of Frankfurt University points out, the locations given by the bee dances serve to guide not only the gatherers of nectar and pollen, but also the water carriers. Furthermore, when bees are swarming, their scouts report on promising nesting places with the same figures from their dance language. In other words, bees use the same symbols in three entirely different situations.

The bees have another way of varying their communications: the duration of the dance. If the scouts are 'enthusiastic' about the new nesting place they have found, they dance continually on the cluster of swarming bees for several hours, again and again indicating the distance and direction of their discovery. But if the bee considers its find only moderately good, it does not venture to dance more than ten seconds. Martin Lindauer[26] comments: 'It is almost ashamed to utter its offer of so wretched a residence.'

Thus there are creatures in the animal world who have the capacity to alter the tone of their communications!

In speaking of little bees, we must take occasion to refute a great philosopher. Friedrich Nietzsche once asserted: 'Animals have no language because they always immediately forget what they wanted to say.' Today Nietzsche would have to reconsider that point. On October 27, 1961, Lindauer's[27] worker bees informed the members of their hive that a new source of food could be found a thousand feet south of the hive. Almost at once, however, a spell of cold wet weather set in, to be followed by winter. On the first warm day in spring, March 30, 1962, 173 days later, the old worker bees reappeared at the same feeding site. Thus the insects had remembered its position for half a year.

Nevertheless, even as we marvel at the language of bearded titmice, ants, or bees, we are bound to recognize the vast gulf between this and human language. In all the examples given above we have variations of instinctively guided reactions. This is especially clear in the case of the titmice. The combinations in sentence structure are really combinations of the birds' emotional states. Each mood is expressed in a special call, and the melody of the song varies in keeping with the succession of moods that the bird experiences.

At this point the analyst seeking to define the difference between man and animal might reach the following conclusion: animals possess only a gamut of instinctively conditioned emotional utterances. We might roughly compare the level of animal language with laughter and weeping, and with those exclamations universally understood by men of all languages, such as Ah!, Oh!, and Ow! On the other hand, animals can never invent sound-patterns of their own impulse, in order to denote other animals or objects. But in saying this last we are once again somewhat precipitate and perhaps off course.

Aside from the fact that this apparent difference between man and animal will possibly be disproved by the dolphins, once we have made further progress in deciphering their language, it was discovered in 1962 that birds can give each other names of their own accord, and can also call each other by name.

Here is a discovery of tremendous importance. This is how it came about.

At the Max Planck Institute for Behavioral Physiology in Bavaria Dr. Eberhard Gwinner[28, 29] kept several ravens. These are probably the most intelligent of all birds, and he wanted among other things to investigate their linguistic talents. For ravens do not only caw. They also imitate foreign noises such as the clatter of storks or the screeching of a circular saw. They can reproduce human words better than parrots. Moreover, ravens have distinct predilections in this matter. For example, one raven named Wotan liked to imitate the barking of dogs, whereas his mate Freya took pleasure in gobbling like a turkey.

One day Wotan had flown off and disappeared. In despair Freya did something she had never done before: she constantly sounded her missing mate's favorite call: barking. And although Dr. Gwinner had previously assumed that Wotan had no idea how to produce his Freya's 'hit song', he now learned otherwise. For Wotan responded with the turkey gobble which he had apparently never before practiced.

Both birds actually understood what was meant. They felt personally addressed; it was as if they were calling each other by name, and by continual calls they found one another. Once they were back together again, each returned to his own favorite 'melody'. From that time forth, Freya did not 'bark'.

Around the same time Johannes Kneutgen[28] at the same institution observed behavior in dayals completely in line with this conclusion. Here, too, a disconsolate bird used its mate's favorite melody to call whenever the mate had stayed away from home for more than an hour. Thus it seems likely that 'calling by name' is not peculiar to individual birds, but is practiced by several species, perhaps by all imitative birds.

For decades science has asked in vain why parrots, parakeets, ravens, mynas, crested larks, and other birds can repeat human words. What kind of survival value can it possibly have for mocking birds to imitate the calls and songs of other species? The garden warbler[30] can sing exactly like a chaffinch. A redstart can sing like a tree creeper and a reed warbler like a willow warbler. Can it be that the imitators merely want to taunt the other birds—like schoolboys?

We know the answer at last. The talent for imitation is not a jest on the part of nature. It has a definite meaning of considerable importance, for it is a device for one mate to call another, so that the partnership will not be broken up because one or the other of a pair loses its way. It is probably not a matter of chance, but one of the marvelous consequences of this talent, that among parrots, parakeets, and corvine birds males and females remain faithful to one another in lifelong monogamy. But at the same time we may also regard this ability as a clue to the origin of a human-type language.

What we have here, moreover, are the first beginnings of something else that philosophers have again hitherto considered to be the sole privilege of man: the phenomenon of tradition. By tradition we mean the passing on of acquired abilities to other members of the species, especially to succeeding generations.

The songs of birds can be taken as an excellent model for such handing down, for in most species the young have to learn singing, or at least parts of the tribal song, from their elders.

Singing instruction begins at very different times among different species of birds. Some birds have what might be called 'natural talent' and have no need of learning; among these are the white throat (*Sylvia communis*) and the blackcap warbler.[31] They are born with an inherited disposition to sing the whole repertory of their baby, territory, mating, autumn, and winter songs. This means that they can sing correctly 'automatically'—even if man hatches them in incubators, and their whole lives long they never hear birds of their kind singing. But it also means that they are slavishly bound to the 'notes' prescribed by their inherited disposition. That is why an English warbler sings exactly the same as a French or German warbler.

The case is quite different for nightingales, blackbirds, chaffinches, dayals, white-crowned sparrows,[32] cardinals, [33] and many other songbirds. All they inherit from their parents is a fundamental motif, a musical theme, a stylistic bent, a range of variation within which they have ample room for individual shaping of their own song. Before they themselves can sing, young chaffinches must learn acoustic tricks and trills from their parents. If they are prevented from hearing models, their singing remains arrested in primitive series of tones throughout their lives.

Johannes Kneutgen[34] has observed young dayals literally 'attending school'. Early in the morning five babies flew out of the nest to a tree fairly close by and perched on either side of their father, who then 'instructed' them. While he sang something to them, they listened attentively with tilted heads. After a while they began to sing along with him, very softly, still listening. The father repeated a melody like a trained pedagogue, continuing until all the children could strike every note correctly and no discords marred the choir.

A musically gifted dayal is also quick to learn melodies played to it from a tape recorder. But very few birds are capable of such abstraction. A cardinal[33] and many other singers[30] must have their fathers while they are young, in order to learn to sing. All efforts by Canadian and German investigators to teach them singing with a tape recorder have failed. These small bundles of fluff must have a feathered mentor; in addition, their singing teacher must have a close personal relationship to them.

Given this factor of learning, it is not surprising to find the phenomenon of dialects in the avian realm. The young reproduce the dialect of their parents. And the parents govern their own singing by what they hear in the immediate vicinity. For if they want to sing a rival out of their territory, they can do so most effectively by singing the rival's notes.

Nature, however, has so arranged matters that the vocabulary of chaffinches (according to the findings of Freiburg zoologist Gerhard Thielcke[35]) allows for at most six dialects. These birds are simply not capable of producing more variations than that. The garden warbler is also limited in the same fashion.

The consequence is rather strange: in the garden warbler population of Germany, one dialect group borders on another in mosaic fashion; there are well-defined lines without transitional areas. North of Freiburg the birds sing Dialect A, south of Freiburg Dialect B. Round about the city we encounter Dialects C through F. But that exhausts the possibilities, for which reason zones with familiar dialects reappear. Thus in the vicinity of the North Sea, in the Ruhr, in Bavaria, and in England and France as well, there are regions where the garden warblers sing the same 'dialect' as their fellows in the Black Forest.

The greater the innate capacity of a species to vary its song, the more numerous and individual are the dialect zones which may be distributed over whole countries. In its most extreme form, this principle is illustrated by the Australian flutebird and a Mexican finch.[36] Among these birds no two males sing the same melody. But each bird remains constantly true to its own dialect, once he has chosen it. Singing in the jungle is a personal means of identification, as we have seen in the case of the black-headed gonoleks (see page 38).

Whinchats[37] are much more concerned about variety. They weave into their specific song various bits of other birds' melodies that they have picked up somewhere. If a whinchat suddenly decides that it likes a blackbird's tune, it incorporates it into its song. The tune then spreads like wildfire among all the other whinchats of a neighborhood, until all of them are singing the blackbird's melody. After a while, however, they tire of it. The birds may then, for example, take up a finch's song, until that enthusiasm also grows stale.

Thus we see the phenomenon of fashion emerging among birds; it appears to spring solely from the desire for variety.

The songs of the bullfinch, however, represent the acme of the 'traditionalistic' impulse among birds. Nestlings always learn only their father's melodies. But a bullfinch when just hatched from the egg regards the first creature he sees moving about the world as his father. This trait can lead to curious complications when, for example, human beings have a canary incubate the eggs of a bullfinch. The bullfinch will regard itself as a canary all its life and will consequently sing like a canary.

The zoologist Dr. Jürgen Nicolai[38] has spent several years investigating this peculiarity. In the course of time many children, grandchildren, and great-grandchildren of the canary-songed bullfinch came into the world and were sold to distant places. But after five years all of them sang exactly like the cock canary who had first taught its song to their great-grandfather.

If a very young bullfinch is raised by man, he regards the human being as his parent. He becomes oriented toward men. Dr. Nicolai whistled melodies to a young bullfinch which then retained the tunes for five years and learned no others. All its sons likewise sang these tunes exclusively, although from infancy on they were able to hear many other bullfinches and some thirty-five other species of bird.

Sometimes this quirk may lead to political embarrassment for the owners of bullfinches. At the beginning of the First World War one such bird learned the German imperial anthem. Four years later, after Germany had lost the war, the Kaiser had fled to Holland, and revolution was shaking the country, the innocent little creature continued cheefully to sing, '*Heil dir im Siegenkranz, Herrscher des Vaterlands,*' and moreover taught the melody to its offspring. The desperate owner covered the bird's cage with thick woolen blankets in an effort to keep the neighbors from hearing the now-outmoded song. Here was one instance in which a tradition in the animal world outlasted even human traditions.

Four-legged Liars

A Canadian baby beaver is credited with a typical bad boy's prank.[39] He belonged to a colony that was fed regularly every morning. Since this thick-pelted rascal always wanted to snatch the best dainties, it was always first at the feeding place.

One day, however, it was late for feeding time, and when it lumbered out of the water all the larger and adult beavers were already gathered around the trough. Thereupon the baby dived back into the river and slapped the water three times with its broad tail. In beaver language that is the alarm signal for extreme danger. Like a flash, all the other beavers vanished from sight in the water, and the cheeky little fellow had the feed trough to itself.

The report goes that he never repeated that trick. Perhaps he was given a thorough thrashing by his duped parents.

This true story is not only amusing. It also refutes the long-held view that animals cannot lie, deceive, or mislead.[40] To be sure, only man can concoct perfected lies. But in the animal world a variety of steps toward lying does occur.

Let us consider for a moment all that goes into this 'achievement' on the part of the small beaver. In the first place, he has to give the alarm signal without having actually been frightened by a predator. He must therefore liberate himself from the fetters of purely instinctual behavior and must combine his action with an intention. That can only be done by a kind of reflection. He must know how his actions affect others if he is to deceive them successfully. Not all animals can do that, by any means. Cunning is a sign of intellectual agility!

As evidence that this is no isolated or chance happening on which we impose an interpretation, we may consider a similar case involving the 'stupid' chicken. A trick practiced quite frequently by barnyard cocks reminds one strikingly of the hoary tale of the shepherd boy who cried wolf. Dr. Erich Baeumer,[41] Germany's leading expert on domestic fowl, describes the trick as follows: When a cock has discovered a rich source of feed at an unexpected place, he coaxes his hens to come to it with loud 'tuck-tuck-tuck' sounds. He surely does this not out of pure charity, but from his duty as a ruler, mingled with a good measure of self-satisfaction. How otherwise are we to interpret the fact that the cock flies into a temper if his 'ladies' do not come promptly? Behind that call, therefore, is the definite intention of issuing a command. And woe to the hen that does not obey his lordship's kindly summons.

Every reasonably intelligent cock—and domestic fowl are by no means stupid—realizes fairly soon that he can summon his hens arbitrarily by using the same call. If, for example, he wants to indulge in love-making, the lazy pasha does not go to the trouble of pursuing his favorite hen. Instead, he posts himself in a place where there is no feed anywhere near, utters feed cries, and waits until all the hens come running. Thus the cock deceives his flock to coax them to him for an entirely different purpose, so that he can pick out the most attractive of his beauties without any trouble.

Another type of deception[42] is habitual in many an animal society. Sometimes a strong cock wishes to administer a beating to a weaker one. The victim is spared if he begins to cackle like a hen when he approaches his powerful enemy. This cry is meant to say approximately: 'I'm not a cock at all, you know, only a poor hen, so let me be.' In most cases this is an effective ruse; the stronger cock is content with his rival's admission of weakness.

Herring gulls pretend helplessness in order to commit their 'crimes'. The Kiel zoologist Professor Adolf Remane[43] reports on this form of animal lying, which is by no means so innocuous as the deceptions of cocks. Significantly, it occurs only in those bird sanctuaries in which the gulls have increased to the point where overpopulation produces degeneration phenomena that shatter the birds' community life.

In such bird colonies a number of herring gulls develop into cunning murderers who eat the chicks of their own species. It is not too easy for them to approach their small victims, since the parents drive away any stranger with slashing blows of their beaks. The infanticide therefore proceeds as follows. The gull wanders casually about the neighborhood of a nest containing chicks. As soon as the angered parents attack it, the adult cannibal behaves like an innocent chick itself. It adopts the typical position of young herring gulls, crouching, drawing in its head, and—the height of impudent slyness—holding its beak up in the begging gesture.

No adult gull can harm a creature in this helpless position of a baby chick. As a result of this hypocritical posture, the predatory gull is allowed to stay around. Sometimes it waits for hours in this position until the two parent gulls leave the nest. Then it darts forward in a flash, seizes one of the chicks, and devours it.

This example shows vividly how animals can 'lie' without the use of words. The intended effect is achieved by gesture language alone.

But is such deception actually intended? Can we really speak of lying in this and similar cases? In order to look more deeply into the matter, the American zoologists Norton, Beran, and Misrahy tested animals with a piece of apparatus ordinarily applied only to men: the lie detector.

Their subject was an opossum, the cat-sized American marsupial. These animals have the habit of falling into a deathlike rigidity when attacked by a predator. Deceived by this behavior, the predator often decides that the opossum is dead and lets it alone.

But playing dead may not necessarily be a trick. It might also be the result of shock, a swoon caused by fright, the numbness of terror.

For the purposes of the experiment an artificial enemy was employed, a wooden dog whose jaw could be opened and closed like the mouth of the crocodile in *Peter Pan*. Simultaneously the scientists played a tape recording of a dog's bark. At first 'bite' the opossum tried to bite back. But then it collapsed, apparently lifeless. Its head drooped, its mouth gaped open, and its eyes stared glassily for some ten minutes.

Oddly enough, the lie detector showed that the brain-wave patterns of the opossum reflected intensified nerve activity only at the beginning of the attack. Afterward, the brain operated quite normally, and as evenly as it would in an awake and not especially disturbed animal. Even when the deathlike rigidity ensued, there was no change in the pattern of the electrical brain activity, and likewise no change upon awakening from the 'sham death'. Contrary to expectations, the brain-current curves characteristic of shock, unconsciousness, or sleep were not registered.

From these facts the scientists concluded that the opossum had not suffered shock. It was only 'playing possum', using a trick to save itself from danger.

A life-and-death deception is employed against the francolin, a partridge that dwells in the African bush with its harem of hens. Its deadliest enemy is a striped cousin of the mongoose,[44] which can imitate the francolin's cry to perfection. This talent facilitates the mongoose's hunting. As soon as it discovers a flock of partridges in the vicinity, it crows like a cock. The cock partridge thinks that a strange francolin wants to fight him for his harem, and rushes toward the sound in blind fury. The mongoose need only slit his throat; his meal has rushed into his mouth.

On the other hand the mongoose itself falls victim to a different trick. Ordinarily it is death on snakes. In the tropics, however, according to Dr. Irenäus Eibl-Eibesfeldt,[45] there are vipers whose tails confusingly resemble their heads. If a mongoose attacks, the viper raises its tail threateningly as if that were its head. The mongoose is deceived, bites the supposed head, and within fractions of a second is itself bitten by the poisonous snake's real head.

Whether the snake is clever enough to use such a feint deliberately, or whether this is an instinctive form of behavior, cannot yet be determined. But both types of animal deception do exist.

Certainly we are dealing with instinctive reaction in the case of a rare phenomenon observed in the Indian Ocean by Professor Heini Hediger,[46] the head of the Zurich Zoo. He saw a large school of tiny fish, each no bigger than a finger. Suddenly a predatory barracuda approached. A quiver ran through the school, and instantly the dwarf fish closed ranks and 'as one man' assumed a formation that looked like an eighteen-foot giant shark. Four times in succession this sea monster composed of thousands of tiny units leaped high in the air as if it were a single organism, then splashed back into the water dolphin-fashion. Seeing its prey so abruptly transformed into a huge shark, the barracuda paused for a moment 'as if it had to wipe its glasses'. Then it fled in terror.

Wherever human beings manifest all-too-human weaknesses, there are other human beings prepared to exploit these weaknesses by means of small deceptions. Amazingly enough, the same situation exists among animals. The zoologist Dr. Otto von Frisch[47] relates one such case. One day he went for a walk by the shore of a lake with his jackdaw, Tobby, and his marten, Fazi. Both these creatures were devoted to their master and jealous of one another. At the lake the marten caught a small fish and wanted to begin eating it. The daw was also fond of fish. It flew close, with signs of intense excitement, perched on a near-by tree, and seemed to be considering how to obtain possession of the dainty.

When Fazi had already devoured the fish's head, Tobby flew over the marten, attracted its attention by loud cries, then hopped into a heap of dry leaves by the edge of the road. There it began pecking enthusiastically, as if the most marvelous things could be found among these leaves. Fazi, an inquisitive animal always worried about missing something, hurriedly ran toward the spot. But as soon as the marten came within six feet of the bird, Tobby pretended to be frightened and soared into the air. And while Fazi sniffed and rummaged among the leaves, the daw snapped up the fish and disappeared into the trees with his booty.

Every dog owner can undoubtedly report similar crimes by his pet. Only in these cases it is not the animal but man who is tricked. In every case a considerable amount of social intelligence (see page 198) is needed for such performances. And it is highly illuminating to see how a meager capacity for communication, coupled with a tiny quantity of egotrism, reflection, and intelligence, leads promptly to the phenomenon of lying.

Monkeys and apes, of course, provide the sterling examples of such behavior. Consider, for instance, the ruse of an ingenious rhesus monkey in New York's Bronx Zoo.

One day this sly fellow vanished from the large, rocky, ape enclosure, and several days passed before he was found in Bronx Park and captured again. The fence, the moat, and the rest of the installation were carefully checked. No escape routes could be discovered. But next morning the runaway was gone again.

Once again a squad of police had to be called in to capture the animal. With the monkey back in the enclosure, an attendant hid nearby to see whether he could detect the monkey's method. At first dawn he at last saw the animal fetch a banana from a hiding place. Evidently the monkey had saved it specially for his escape plan. Banana in hand, he ran to the broad moat that bordered the moose range and swung the banana back and forth—exactly like a scientist who uses a reward of food to persuade an experimental animal to perform.

Sure enough, a large moose soon came swimming up to the monkey. The clever but water-shy monkey thrust the banana into the moose's mouth, as a kind of ticket, leaped on its broad back, and rode this 'ferry' to the adjacent enclosure. From there the monkey had no trouble at all escaping.

But the weirdest behavior is found in the harems of the baboons.[48] In the open-air enclosure of one zoo the strongest male established himself as sultan and denied all other males access to his females. He would not even tolerate the slightest flirtation. But the ruler of the enclosure could not be everywhere at once. If he happened to be taking a nap in the shadow of a rock, the females promptly tried to cheat on him. One lady of the harem, who had been neglected by the sultan for some time, seized such a favorable opportunity to display all her charms brazenly, in order to arouse one of the bachelors.

At just this moment the sultan reappeared, and a fantastic comedy began. As if she were being molested against her will, the adulterous female tore away, gave the male she had just been wooing a slap, and flew with loud wails into the arms of the astonished sultan, 'complaining' to him by throwing looks at her potential lover, making throaty sounds of fury, and drumming the ground with her forearms. She achieved her purpose. The sultan, who ordinarily punished only the female for such adulteries, believed this artful lie. First he gave the innocent bachelor a thorough beating, then heaped caresses upon his 'sorely offended' wife.

3. Love—Fidelity—Love of Neighbor

Ardent Love at −60°

Penguins—to the zoo visitor they look like frock-coated caricatures of men. We laugh at their waddling and their dignity. But the scientists who have studied them closely in the Antarctic during the past few years no longer find them humorous. They have found them—at temperatures of −60° and amid blizzards with wind velocities up to ninety miles an hour, in the darkness of the night below the Antarctic Circle—the most self-sacrificing and loyal creatures imaginable.[1, 2, 3]

It is the middle of March. The Antarctic summer is nearing its end. The sun scarcely appears above the horizon any more; the belt of ice surrounding the white continent is piled up into ramparts by storms; and the sea is gradually freezing over. But before it does, a great variety of life fills the chilly water between the floes. Here and there flocks of emperor penguins, looking like playful dolphins in the water, head purposefully for their time-honored assembly grounds.

They come from great distances, these swimming migratory birds of the Southern Hemisphere. From the direction of the Atlantic, the Pacific, and the Indian oceans they have put several thousand miles behind them. No man knows their exact route, and it is a mystery how year after year at the appointed time they reach a great ice floe at the margin of the mainland ice at more or less the same spot they touched the year before.

While all other migratory birds move to warmer parts of the earth before the descent of winter, the emperor penguins seek the coldest, darkest, most hostile zones of the earth at precisely the coldest season of the year. There they mate, brood, and raise their young.

The members of the first flock to arrive seek out an especially large ice floe, in the center of which they are relatively safe from predators, especially seals and sea leopards. At first the floes are still thin and when a new flock of some two hundred birds arrives, the floating saucer often breaks apart under their weight. Then the entire company has to look around for a new and sturdier floe.

Finally some six thousand loquacious emperor penguins are assembled. It must be startling to come suddenly on such a crowd in the icy wasteland. As the American biologist Dr. E. Pryor of Ohio University has discovered in some two years of research in Antarctica, there are a total of twenty-one such penguin assembly sites around the icy continent.

When new arrivals approach the settlement, they perform a 'greeting ceremony'. Dr. J. Prévost,[1] the French polar scientist, describes the ritual as follows: 'The penguin raises its head, extends its neck, and then rubs its "ears" against its flippers. Then the bird slowly bends its head to the ground, takes a deep breath, and begins to sing. As soon as it stops, it raises its head gently and listens. Then, as if it had received permission to enter, it waddles at a leisurely pace through the groups of penguins standing about, continuing to sing.'

Soon afterwards the courtships begin, in frock-coated dignity. With beaks held toward the sky, the males steadily sound their courtship song.

For the great majority of the penguins, the song is directed at the male's wife of the previous year. For almost all penguins practice monogamy. In the case of the gray penguin Dr. L. E. Richdale[4] has determined that couples stayed faithfully together for eleven years— though only during the nine-month brooding period. For in the Antarctic spring, when the crowd of six thousand breaks up into smaller flocks, husband and wife usually travel to distant feeding grounds in separate groups—much as do our gulls.

But how is Mr. Penguin to find his wife in the midst of this close-packed throng of many thousands of birds? One bird looks like another—and for once this is even more true for the eyes of penguins than for the human observer. For penguins are so nearsighted (on land, not in the water) that they can scarcely distinguish an egg from a stone at their feet, and at a distance of ten feet they do not know the difference between a kneeling human being and a fellow penguin—which after all stands four feet high. Indeed, what need have they of good sight on land? For during the Antarctic winter it is usually so dark that you cannot see your hand before your face.

Penguins therefore do not judge by appearance, but by the song. Singing without a pause, sometimes screeching desperately, the male wanders back and forth amid the noisy throng. As soon as his previous year's wife recognizes him by his voice, she joins his song and moves, in a stately manner, in his direction.

Dr. Prévost describes the joy of the reunion: 'Then the two stand facing one another, necks bent far forward and heads tilted amiably to the side. Suddenly they throw their heads back, tip forward slightly and lean against each other, breast to breast. They stand this way immovable for some time, seemingly unaware of everything that is going on around them'. The two penguins appear to be in seventh heaven.

A bachelor looking for a wife has a relatively easy time of it. Quite often two single females quarrel over him. They join him of their own accord; each showing her brood pouch to prove that it is superior; and the two females bicker incessantly. The male acts bored and seemingly takes no notice of the screeching 'ladies'. This spectacle can go on for a day or two, until finally one of the two applicants voluntarily withdraws. There is no bigamy in Antarctica.

It is likewise beneath the dignity of a male penguin to quarrel with another male over a female. In general the use of physical violence among penguins is very rare. Extremely outraged penguins confront their opponents in the spirit of a massive bar-keeper descending on an obstreperous patron. Then they abruptly thrust their bellies forward against the antagonist, so that he falls backward. With that, the dispute is usually over. Bloody battles with blows from beaks virtually never occur among emperor penguins.

Occasionally, however, vigorous swipes are administered. Thus when Dr. Rivolier, the physician of Dr. Prévost's expedition, wanted to take a penguin's temperature, he received a slap that 'knocked him down to the count of nine'.

The 'birth' of a single egg—there seems scarcely any other word for the process—takes place some two months after the landing, during

A pair of emperor penguins greeting each other after a long separation.

which time the birds have been fasting, and has some almost human aspects. The female doubles up from severe pains, while the male keeps running agitatedly around his wife, desperately eager to help and perplexedly unable to do so. He meekly puts up with a succession of blows from her beak. As soon as the egg is laid, the male utters a cry of triumph in which his wife joins, though with somewhat less force.

At first she stows the egg in her brood pouch. But after only a few minutes the male insists on taking over the egg. He voices a special 'egg song', and makes innumerable 'Japanese' bows. Then, with much groaning and trembling, the juggler's stunt is performed. Afterwards, both begin a new song. And now the female waddles around her husband. Slowly, she moves away from him, returns, sings heart-rending songs of parting, and parades about stiff-legged. This performance is repeated several times, but each time the female moves farther away, until at last she vanishes in the darkness of the polar night, leaving the brooding male for two full months to the fury of the elements.

While the hungry females set out on their long journey, walking or sliding on their bellies and propelling themselves through the thick snow by rowing motions, in order to fish in ice-free waters, the life of the males who remain behind for two months without food is anything but enviable. In cold which sometimes descends to −85°, amid furious snowstorms, they endure still another two months without a crumb of food. When a storm breaks—and in the Antarctic winter there is scarcely a day without a storm—driving inconceivable masses of snow before it, the 'pregnant' fathers press together against one another forming a kind of carousel that revolves slowly in a circle. To use the technical term, they make a 'tortoise'.

Before the Antarctic blizzard descends, 500–600 emperor penguins form a 'tortoise' for mutual protection.

This circular motion obviously arises because the birds most exposed to the wind slowly move toward the protected side of the group. Moreover, those standing in a position where they are shielded from the wind are apparently willing to expose their bodies to the worst of the storm for a while, for the good of the others. Incidentally, the carousel does not quite have a rigid axis. After a blizzard that raged for two days Dr. Prévost measured a displacement of 660 feet to one side. As soon as the storm subsides, the 'tortoises' break up.

Without this sense of community, penguins would not survive the Antarctic winter. That was impressively demonstrated by the French zoologist Dr. Sapin-Jalustre and his colleagues. In spite of darkness, cold, and storm, the scientists forced their way into the middle of the 'tortoises' and measured the body temperatures of the birds. They then took several marked penguins out of the group for weighing.

They found that a penguin standing alone in the icy wasteland loses as much heat in a storm as he would under windless conditions at an outside temperature of − 292°—a horrifying notion. If the bird is part of the 'tortoise' he loses 100 grams of his cushion of fat every day that the storm goes on. But should he be exposed to the forces of nature alone, he would lose exactly twice as much weight. After a total starvation period of four months the males, which initially weighed some sixty-six pounds, have lost almost half of their body weight. Without mutual aid there would be virtually nothing left of them.

How eager a brooding male penguin is to leave this icy inferno and go after some food, like the females, is apparent from his behavior with a lifeless egg. The eggs of some of the birds prove to be infertile. And as soon as the penguin realizes that there is no prospect of his egg's hatching out, he throws it away and makes off at once—toward the fishing grounds. But the others remain faithfully dedicated to their task.

Shortly before the return of the females, the males experience a happy event. Inside the warm belly pouch a naked baby penguin the size of a roasting chicken hatches from the egg. Emaciated as the male is, it lovingly feeds its hungry progeny. But with what? From its crop it brings up a milklike secretion. These male birds actually produce 'milk' for their offspring. When the survival of the species is at stake, nature comes forth with prodigies.

At last, at the beginning of June, the female returns, having eaten her fill and bearing seven pounds of fish in her gullet as a gift for her husband. The reunion is once again accompanied by displays of emotion: bowing rituals of Far Eastern courtesy, piercing singing, leaning against one another, and 'embracing' each other with their wings. But above all the pair gaze with common pride upon the child that the father has meanwhile hatched.

During the rearing of their young one, which takes a surprisingly long time, from June to the beginning of December, the parents spell each other for intervals of three to four weeks. Only one thing will keep them from returning to their duty: death. At the beginning of September the warming rays of the spring sun break up the ice belt around Antarctica into huge floes. That is the time for young and old in the realm of the emperor penguins to leave the colony in small flocks, to seek their hunter's luck in the expanses of the ocean and build up their strength after the exertions of marriage and child rearing.

It is hardly surprising that the French scientists were fascinated by the human not to say superhuman marital fidelity of the penguins.

This astonishing phenomenon, which goes so far beyond sexual pairing, must also be traced in the life of another bird, the common raven.

Ravens Prefer Engagements

The story of this hopeless love might be taken from a cheap novel. He courted her according to all the rules. But she disdained and humiliated him, for she loved another and threw herself at this other in the most awkward way. Alas, insuperable 'class differences' stood in the way of their union. Nothing very odd about such a triangle—except that in our case the rejected suitor was a raven by the name of Goliath, the adored sweetheart a female raven named Davida, and the mate for whom she swooned in the confusion of her affections was a . . . human being, the behavioral scientist Dr. Eberhard Gwinner.

Such grotesque emotional turmoil arises in many birds whenever the person who tends them raises them like a real mother bird from their very first day of life. Such birds actually regard the human being as their mother. Later they often try to win him as their mate.

In this way Dr. Gwinner[5, 6] at the Max Planck Institute for Behavioral Physiology was able to play the part of a raven among ravens and make some extraordinary discoveries about the social life of these highly intelligent birds—which, because of their very intelligence, no doubt, manifest such astonishingly human traits.

The male raven brings his bride (right) a gift of food. Among older marital partners this feeding ritual at mating time can be done symbolically, without the offering of actual food. (After E. Gwinner.)

The rejected Goliath did not accept defeat easily. Obeying one of the most ancient principles of courtship, he offered the lady of his heart small gifts. Quietly, he prepared tasty morsels of meat and hid them under the bark of trees.[7] As soon as he saw his sweetheart, he would fetch out one of these dainties and march up to her with stiff steps, twitching his wings and uttering the coaxing cries that signify food.

Normally, when a female raven responds to a male raven's suit, she squats, flutters her own wings, whimpers, and takes the proffered gift. She understands the gift as a symbolic statement that the raven is willing 'to support a family'.

Davida, however, always refused to accept the gift. Gradually the desperate suitor became more and more importunate. At last, with loud shrieks, he tried to feed her by force. But she took off. Therefore Goliath flew after her, screeching, and pecked furiously at the female. Naturally this behavior scarcely improved her attitude toward him. In periods of total depression poor Goliath sought a surrogate for his feelings and cawed his mating call to passing airplanes.

This situation went on for half a year, until the following spring. Every time Davida lovingly ruffled the zoologist's hair with her big beak, and Dr. Gwinner responded by caressing her plumage with his finger, the male raven attacked his human rival. But Davida always defended the scientist vigorously, and successfully, from the dangerous blows of the jealous male's beak.

One day the female actually flew into the nest that Goliath, with hope springing eternal, had built for her. She lured her human friend to the place and coaxed him, by uttering the proper calls in raven language, to help her improve the nest.[8] Indefatigably, she brought more and more twigs, which Dr. Gwinner had to weave in carefully.

In spite of this inauspicious beginning, the story had a happy ending after all. The zoologist had to leave for a trip of several weeks, and when he returned he saw Davida and Goliath in the midst of the prettiest mating ritual. At maximum speed the male raven shot through the air performing artistic feats of acrobatic flying, while Davida sat on the nest watching. Then both took wing, flying in formation. Goliath followed diagonally behind Davida and reproduced her every loop and twist as if he were tied to her. No wonder the scientist could no longer compete.

The drama of Davida and Goliath demonstrates the pre-eminent importance of the 'engagement' in the lives of ravens. Of course this engagement must not be understood in the human sense—although it comes to much the same thing—for the birds are compelled by nature to undertake it. They cannot do otherwise.

Mating between ravens is biologically possible only for a few days in spring, after the birds are two years old. But a full year earlier male and female feel a compulsion to pair off. This has nothing to do with sexuality, incidentally. To this day there is a widespread view that all social behavior may be reduced to the 'instincts' of sexuality, escape from danger, and aggression. But this old theory is totally mistaken. Helga Fischer[9] of the Max Planck Institute for Behavioral Physiology has asserted emphatically in connection with graylag geese: 'Their association is founded on a special instinct—the associative instinct.' There is a high probability that this same instinct is operative in monkeys and apes.

Sexual attraction can bring animals together only for minutes, sometimes only for seconds. If there is to be any longer period of common life, with all its considerable advantages, a different and wholly independent impulse must control it. In ravens this instinct for attachment emerges at the age of one year, whereas the sexual instinct stirs only a year later. The birds must therefore willy-nilly go through approximately a year of 'platonic' engagement before they can consummate the marital bond.

The purpose of engagement among ravens is much the same as it is among human beings. The more intelligent an organism is, the more pairing functions independently of the pure gratification of the sexual instinct and the greater the part played in mating by mutual liking and the sense of a common bond.

Ravens are strictly monogamous. Once paired, they practice lifelong fidelity. Since they live to be almost as old as human beings, but 'marry' at the early age of barely two, they remain linked to one another longer than any human couple.

During the engagement period ravens may still change partners, and they do so as often as necessary until they have found their proper mate. Sometimes female ravens let their suitors dangle for a fairly long time—as Davida did her Goliath. Usually, however, engaged couples assist each other through all dangers. Together, they combat intruders and are therefore, as a pair, safe from attack by single ravens. They also fly together to hunt for food, or play with one another and tenderly scratch each other's neck feathers with their beaks.

A bit of flirtation with strangers is quite permissible for either member of an engaged couple. By the rules of propriety among ravens, other suitors may also do a little scratching of the bride's neck feathers. But from the day of sexual congress on, such caresses are strictly forbidden. Dr. Gustav Kramer[10] once observed a newly mated female being caressed ever so little by a strange lover, while she was sitting in her nest. At that moment the husband arrived home. In a towering fury he flew at the intruder and drove him off, not returning himself until late in the evening. The Casanova was never seen again in the vicinity. He had evidently been chased clear out of the region.

In spite of this exemplary fidelity, raven marriages are not without their complications. Davida, for example, was physically much stronger than Goliath. But she did not make use of her superiority. In minor conflicts, as to whether the nest should be built here or there, whether they should fly to one place or another, she always accepted the subordinate role. Only when it was a question of food or the best sleeping place did she show who was master in the house.

The female raven Eva, newly married to her Adam, behaved quite differently. From the first day of their marriage she took every advantage of her feminine prerogatives. Adam slaved all day long to bring twigs for weaving into a nest. By raven custom Eva should have helped with this work. But for a whole week she sat lazily on a distant branch, merely looking on. Then suddenly, with the basic structure of the nest already finished, she made it clear, by a steady series of coaxing calls, that she wanted the nest where she was perched instead. So Adam began his building all over again.

All went well for a while, but not for more than a week. Then Eva abruptly decided on the first nest, and Adam had to complete construction on this dwelling. No sooner had he finished than his wife took it into her head to move to the second nest. Finally Adam took to flying directly behind Eva with a twig in his beak. Back and forth he flew, from one nest to the other, until she had indicated her preference. But finally he grew tired of being humiliated and led around by the beak. Paying no further attention to Eva, he worked only on the first nest, and in the end the female raven had no choice but to move in there.

Human beings in such circumstances would be on the verge of a divorce, or already over the brink, but that was out of the question for Adam. Oddly enough, divorce virtually never occurs among ravens.

Male and female remain faithful all their lives, even if they are quarrel-some and a scourge to each other.

It must be granted, however, that not enough research has been done on raven marriages for us to be able to state with assurance that this is always the case. Among ravens, as among graylag geese, there may also be cases of homosexuality, adultery, courtesanship, family rows and divorces.

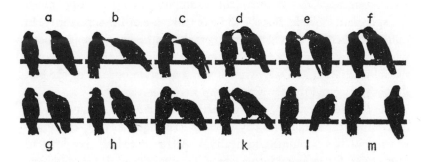

Vain attempt by male raven to approach a female (left): (a) *May I come a little closer?* (b) *Scratch your feathers a little?* (c) *I am your humble servant.* (d) *How dare you, sir?* (e) *I only wanted to scratch you a little.* (f) *But you're getting fresh.* (g) *All right, I'll behave.* (h) *But don't you want to scratch me once?* (i) *Here, in my neck feathers.* (k) *I see, here!* (l) *Nothing doing, I guess.* (m) *Good-bye.*

Among ravens, scratching is a gesture of goodwill that has no sexual overtones. (After E. Gwinner.)

The expert can recognize by the condition of the nest whether or not a raven couple is leading a harmonious marriage, whether the two have artfully done the weaving together, whether the nest is a crude, shaggy masculine project or whether it has been constructed by the female alone, with lighter material. For all such styles of raven nests are found.

During their year of engagement the adolescent ravens often spend hours every day practicing nest building as a form of play. The way they transport building materials through the air to the site of the nest is their most brilliant achievement. Less intelligent birds can carry only one twig at a time. Otherwise they have the same difficulty as raven beginners who enthusiastically take far too many twigs into their beaks and lose them one by one during the flight.

An experienced raven, however, performs an astounding intellectual feat. First he balances a twig in his beak until he has found the center of gravity. He observes where this is, puts down the twig, and repeats the test with a second, third, and fourth twig. Then he places his sticks

After the raven has first tested the center of gravity of each twig, he can carry several through the air simultaneously.

side by side, matching all the centers of gravity, grasps the bundle at this point, and flies debonairly away with it.

Brooding is the business of the female alone. Once, however, the Danish ornithologist Dr. L. Moesgaard[11] observed an astonishing exception to this rule. During the last week of brooding, the female raven was killed by an owl. The male raven thereupon sat on the five blue-green eggs and continued brooding them. The widower went hungry until his chicks hatched. Then he fed and provided for them in the tenderest manner for about a hundred days, until the nestlings were able to take care of themselves. He must surely have watched the process in previous years and learned how the mother handled the whole affair. In German, cruel and heartless parents are called 'raven parents'. What a misnomer!

Once the half-grown ravens are out of the nest, they join up with other ravens their own age in juvenile flocks. They remain in such flocks until their 'wedding'. Their principal activity during this period is play.[12] In creative play young ravens develop a degree of imagination and virtuosity exceeding anything we have hitherto known in the realm of animals.

In addition to nest building, pursuit games are popular. When one young raven seems disinclined to join, the other who wants to play provokes him. He stalks over to his sluggish companion with an earthworm in his beak, and with a courteous bow lays this morsel at his feet. Fractions of a second before the second raven can seize this supposed gift, the challenger snaps it up under his beak and flies off with it—and, naturally, is furiously pursued by the victim of the trick.

Ravens enjoy acrobatic exercises. Training begins when a bird perches on a thin branch and then falls forward or backward. He lets himself drop and flies off. More advanced 'students' vary the trick by perching on only one leg. The boss of the flock executes what is actually a gymnast's giant swing: once he is hanging head downward, he raises himself to the perch again with a single beat of his wings. When the other ravens try to imitate this stunt, he carries it a step further in difficulty. He then performs on a thin, flexible switch that whips up and down a foot or two.

Among the mass sports is the art of balancing. Starting at the stoutest part of a branch, raven follows raven until the branch becomes thinner and thinner. One after another, the birds fall off. The winner is the one who ventures out the farthest.

Soon the boss tires of this game. He therefore invents a new sport. Picking up a chicken bone, he flies to a branch as thick as a human arm, places the bone on it, and tries to stand on the bone and maintain his balance. For days a whole flock of ravens could think of nothing but practicing this stunt. They were training in self-invented circus acts!

A new fad became established one day when the boss discovered a large, smooth plastic slab in a forest clearing. At first the curious raven slid badly when it landed. Immediately it turned this disgrace into a virtue: it flew at the slab as if it were an airport runway, made a sliding landing, and repeated the stunt again and again until it could land without falling over. It also practiced a technique that is part of the training program of every pilot: coming in for a landing and then 'stepping on the gas' and zooming up to start a new landing maneuver. Fairly soon all the ravens had joined in the fun. Like a wing of fighter planes, one after the other came diving down and made a sliding landing. At first there was a wild knot of birds falling all over one another. But after a few days of practice the formation landed beautifully

One aspect of this mania for play is that the individual raven can thereby raise his standing within the group. Thus it is child's play, in the literal sense, for ravens to acquire rank, and eliminates any need of fighting with one another. In addition, such games increase skills in the struggle for survival.

Here is a single example of that. One summer morning agitated cries of *brrruai* arose at the edge of a grain field. That is the ravens' call of maximum danger. The boss had discovered a weasel, one of the worst enemies of ravens. It had to be expelled. But how could the ravens achieve that—dealing with an animal whose reaction speed is proverbial?

In response to these cries of alarm all the ravens of the juvenile flock came flying up and perched at a discreet distance from the weasel. Then the fun began. One raven alighted in front of the weasel at a distance that seemed suicidally close. As it sprang, the raven flew up in a flash, while from behind a second drove its beak into the weasel's rear. Infuriated, the animal whirled around. But in the same instant a third raven slashed it in the back.

The weasel leaped yards into the air, but invariably clawed at empty space, and as it fell would instantly be struck from behind. Finally the boss fluttered to within sixteen inches of the weasel's nose and lured it into giving chase. Flying close to the ground, the raven moved in tighter and tighter spirals, until the predator was out of breath and dizzy. Thereafter the weasel vanished amid the grain and was not seen again in the neighborhood.

How to Become a Boss

It is time to explain what we mean by the 'boss' of a flock of young ravens. 'Boss' is precisely the right term, for a group of ravens without its leader is just as helpless as a company without its boss.

In this connection, too, we must refute a widespread but incorrect view. The 'strongest' member of an animal community is generally regarded as a tyrant. The thought is that he suppresses all others by sheer physical force, solely in order to obtain the best food, the best sleeping place, and the best female (unless he commandeers all the females). Those human beings who feel they must justify their role as superiors by this hypothetic 'law of nature' merely prove they have no inkling of what nature really is. For the boss in animal societies has numerous difficult responsibilities toward his flock, in addition to his rights. And if he does not meet those obligations, he is promptly deposed.

In the first place the boss is the stable and reassuring axis of his group. An instructive example of this fact also throws light on the character of panic within a human crowd.

The boss of a raven flock had been killed by a poacher, and now the rest of the birds were standing around, anxious, insecure, and unenterprising, in a newly plowed field. Suddenly a stone tipped over in a furrow. The raven standing closest to it gave a slight start. His neighbors were even more frightened by his alarm, and they infected others with their fear. Within seconds the reaction had swelled like an avalanche. The whole flock scattered in wild flight—including the first bird which had been only mildly startled and of course had no idea that the rolling stone, in itself no cause for terror, had been the source of the chain reaction.

If there is an experienced boss with the flock, panics of this sort are averted. The first act of his frightened followers is to look over at him, and if he shows no signs of uneasiness, the birds worry no more but go back to whatever they were doing. The boss is a fount of total reassurance.

In addition, the boss is the conceiver of ideas, the initiator of ventures, the discoverer, the most courageous of all. Whether a new landing site is to be scouted, a strange object or an unknown animal to be investigated for possible tastiness or possible danger—the boss is the first to act. His curiosity does not rest until he has taken the measure of every single unknown thing within his realm.

Once, when the scientist hoped to lure his ravens into their cage by baiting them with all kinds of dainties but found this to no avail, for the clever birds saw through the trick, he employed his knowledge of their psychology. He placed his camera in the cage. The birds had often seen him with this peculiar and obviously desirable apparatus, but had never been allowed to investigate it closely before. The moment the camera was left alone in the cage, the whole raven flock flew in, and the zoologist had them where he wanted them.

This craving for novelty, which is even more strongly marked in ravens than in chimpanzees, is of great service to the birds. Like that supremely curious creature, man, the curious raven seeks out whatever is edible in every environment. Thanks to this quality the raven can adapt to a wider variety of territories than almost any other species. On bird islands ravens feed like gulls. In the African plains they sail at great heights like vultures, on the lookout for dying game. In agricultural areas of North America, Europe, and Asia they subsist on mice and insects. All such sources of nourishment were at one time or another discovered by clever bosses.

A boss can swiftly lose his leading position by incompetence in action. Dr. Gwinner[5] tells the story of a boss whose fiancée died of some disease. The shock of bereavement was so strong and lasting that the bird no longer had an interest in anything. He roosted mournfully in dark thickets and kept away from all his fellows. He simply ceased to do any of the things that were expected of him. Thus he apathetically gave up his leading position, without a fight, to a successor.

In another case the second in rank repeatedly outranked the boss in all sorts of games of skill, and therefore prepared to overthrow him in a bitter struggle. Like fighting cocks, the two leaped up at each other and struck out with beaks and feet. Then they gripped each other's feet and tried to hack like battling knights, each using one wing as a shield to parry the opponent's blows.

In a fight ravens aim their beaks at the joint of the wing, in order to impair the opponent's ability to fly. They never strike at the eyes. The proverbial saying is true: no raven hacks out another raven's eyes.

The fight ends fair and square, on a sporting note. The raven who realizes that he no longer has any chance, surrenders. As the German ornithologist Dr. Johannes Gothe[13] has observed, this surrender obeys the following rule: the loser assumes the position of a raven chick that wants its mother to feed it—begging with gaping beak and uttering childishly yearning sounds. As soon as he does this, the loser is spared any further punishment. But he has forfeited his leadership. Usually his prestige in the flock suffers so greatly as a consequence of his defeat that he is degraded not to second in rank, but to last. This shows how strongly psychic factors influence social life among animals—often far more significantly than sheer physical strength.

In one case observed by scientists, change in leadership had a bitter sequel. The female Cleopatra had had her eye on Nero when he was still boss of the flock of juvenile ravens. After he was defeated, her love for the beaten bird quickly faded away. The ambitious female at once began to fawn upon the new boss. She was successful in her 'ladies' choice' and thus retained the rank of 'first lady', since among ravens the wife has the same status as her husband.

Comparisons with similar modes of behavior among men force themselves upon our attention. Is this because we are anthropomorphizing—conferring human traits on animals? Or is it rather just the reverse?

In any case, it is amusing to see how the new raven boss promptly lays claim to all his rights, but as yet 'knows' nothing about his duties. Consequently, the flock of ravens does virtually nothing. All the birds lounge about purposelessly. Slowly, they grow nervous. The first quarrels spring up among them. These are sharp evidences that the birds are expecting some leadership from their new boss, and that discontent is growing.

The new boss usually finds his first clue to activity in arbitrating quarrels. Now and later it is one of his tasks to see to peace in the flock, at least among the males. As soon as the feathers begin to fly between

13 *At maximum speed the female raven shot through the air, swerved over on her back, closed both wings, and let herself fall a hundred feet. After three caws she checked her fall. The male had to imitate each of her stunts as a test of his fitness to be her mate.*

14 (ABOVE) *The boss raven has found a dead rabbit and adopts the threatening posture to fend off any other claimants. When he returns to his mate he brings her a dainty morsel as a gift.*

15 (BELOW) *But in the later years of marriage he merely feeds her symbolically.*

16 A mighty African bull elephant. He has been known to practice mercy killing upon a fatally ill cow in his herd.

17 (ABOVE) *In union there is strength. Here gnus form a defensive front to protect a newborn gnu from hyenas.*

18 (BELOW) *Loveplay at the foot of Kilimanjaro. The cow elephant (right) holds a stick in her trunk. She uses it as a scratching instrument.*

19 Gannets practice sexual continence in order to prevent disastrous over-population. This colony on the northern coast of Brittany is inhabited by the fortunate birds that are permitted marriage and offspring. Around them is an invisible borderline; beyond it the birds must go unmated.

20 (ABOVE) *Wolves* (left) *are traditionally feared for their ruthlessness. Nevertheless, they can manifest pity for their own kind. A hyena* (right), *too, practices friendly behavior toward all members of its clan.*

21 (BELOW) *When rabbits multiply 'like rabbits', they become a plague. But they too impose birth controls on themselves.*

22 *A murderous bite? No. The lion is only saying to his mate 'I love you so much I could eat you!' This gesture, merely an affectionate nip at the nape of the neck, is part of the mating ritual.*

23 (LEFT) *Lionesses are affectionate mothers. Here the baby is given a thorough bath.*

24 (BELOW) *But lionesses are also capricious. This one has just been ingratiating herself with her mate. But when he approaches her, she starts a quarrel.*

any two, he intervenes. What is more, the boss almost always helps the weaker bird. That need not be an expression of morality. Probably it is purely utilitarian: the stronger of the two brawlers is the more dangerous as far as the boss is concerned—and hence must be subdued.

The boss never takes action against quarrelsome females, however. Hence the females are constantly bickering with one another.

As Dr. Gwinner[5] puts it, the novice boss of a raven flock is gradually trained for his leadership duties by the needs of his 'people'. Slowly, he becomes acquainted with his various duties.

It is by no means only among ravens that the boss has tasks to undertake on behalf of the community. There are similar demands upon leaders in many other animal societies. The self-sacrificing courage with which a chimpanzee boss covers the retreat of his band and the removal of an injured member has already been described on page 12. We shall later speak of the intrepid way a zebra stallion defends his mares and foals against predators (page 135).

But in every case—and here we have a remarkable difference from many a leader of human groups—there is never any question that the lead animal in dangerous situations will expect another to do the fighting for him. The weal or woe of those who recognize his leadership depends solely on his strength, skill, cunning, and experience.

A particularly dramatic example of such behavior has been reported by Professor Irven DeVore,[14] the American anthropologist.

In the East African plains a baboon band of some thirty animals had reached the shore of a small lake and was serenely drinking the refreshing water. All at once the boss baboon noticed that a 'skirmish line' of five powerful lionesses had encircled the baboons. They were hiding in the tall grass and invisibly working their way forward. The fate of a number of baboons seemed already sealed.

The baboon boss took charge with a cleverness that would have done honor to an army lieutenant with years of service. Uttering piercing cries, he drove his frightened, clustering band away from the shore and into the tall grass. Here the invisible lions could no longer see their prey. But it was still too risky for the apes to flee, because they could easily run head-on into a lioness. The leader therefore made his band wait while he crept alone through the grass, scouting. The anthropologist observed the whole spectacle from a tree at some remove from the baboons.

Like an Indian scout reconnoitering around the camp of his enemies, the baboon carefully spied out the position of two adjacent lions without being observed by them. Then he crept back to his band and silently led them in single file, crawling over the ground, right between the two lions without attracting the attention of either of the dangerous beasts. As soon as they reached safety, the baboons scurried up three tall trees and immediately burst into screeches of triumph. Astonished, the lionesses rose out of the grass and stared as if they had seen ghosts.

One particularly remarkable aspect of this affair is the fact that the hero of the day was by no means the strongest baboon in the band. But he was in all likelihood the oldest, most experienced, and cleverest. Thus baboons demonstrate the value of age. They give the lie to the erroneous view that in the animal world the continuance of life is no longer biologically useful as soon as an organism has produced sufficient offspring. Using experience, the abilities acquired over many years, to further the preservation of the species—that is, in itself, a decisive step forward on the long road to *Homo sapiens*. Nor is it a step first achieved by man; it appears independently in many highly organized animal societies and conspicuously in the line that leads to man, that is, among the primates.

Animals Are More Democratic than Men

The scene was the dense mountain forest on the islet of Koshima in southern Japan.[15] There lived a band of some three hundred red-faced macaques, sometimes known as Japanese monkeys. These animals, which are closely related to the rhesus monkey, also had a boss, one rather advanced in age. One day he was attacked by a solitary young male monkey. There was a long, merciless fight, in the course of which the boss was injured and finally worsted.

But then something happened which may not so greatly surprise the reader who has reflected on the nature of 'bossism' among ravens and baboons. The Japanese zoologist Dr. Mizuhara[16] observed the remarkable behavior of the females of the monkey band. They remained loyal to their old, defeated, beaten, bleeding, and limping boss and simply refused to recognize the new ruler. One might almost say: they went on a general strike. All the young dictator's efforts to beat the group of females into submission failed in the face of the whole band's passive resistance. At last the victor had no choice but to remove himself from the scene and leave the old boss in his position of leadership without further fighting.

Here, truly, was a 'democratic plebiscite' among animals!

If, however, we consider these events more closely, they appear somewhat less 'human' than at first sight. All behavior among animals is directed toward survival in the struggle for life. But what use would it be to an animal society if it blindly allowed every muscular idiot that came along to issue orders that might lead it into perdition? For the sake of survival the members of many gregarious species of animals are therefore more democratic than many a human society.

The prairie dogs of North America, for example, carry on a form of electioneering. These mammals live in the Great Plains in 'towns' that may contain hundreds or thousands of inhabitants. In the course of a year's observation of them in the wild, Dr. John A. King[17] discovered that every 'clan', consisting of twenty to thirty animals, strictly defends its part of the town against its neighbors. The entire family group immediately attacks an undesirable visitor and throws it out of its area.

There is a reason behind this. As soon as one generation of prairie dogs has grown up, several bold young males move out to conquer a part of the town, including its inhabitants, for themselves. Mere muscle power—biting and attacking the old inhabitants of the area— does not do the trick. For among these animals, too, the entire clan of the conquered territory may remain faithful to their old boss and leave their settlement with him. They prefer a refugee existence with all its troubles and dangers, would rather dig new burrows in still uninhabited outer areas, than obey a tyrant they do not like.

Weeks before the decisive struggle, therefore, the candidate must carry on an election campaign in the enemy territory—an undertaking that makes the highest demands upon his diplomacy, empathy, and boldness. The telling factor is, above all, persistence. The policy of the successful prairie dog is simple: to stay as long as possible in enemy territory. He lets himself be driven off by the clan without resistance, but immediately comes into the area again at some other spot. In all these actions he makes a point of moving as slowly as possible, of showing no exaggerated fear, and of behaving as inconspicuously as possible, trying not to attract attention and pretending that he already belongs.

Another aspect of the prairie dog candidate's electioneering: he exchanges a friendly kiss with one of his future wives.

After a while the candidate cautiously approaches females who happen to be alone and tries to play with them and exchange caresses. These are, as it were, election promises. The young prairie dog must wait until he has made friends with most of the enemy boss's group and has won some popular backing. Only then can he venture to fight.

An incessant plebiscite is taking place in every flying flock of birds. Crows, for example, are constantly deciding anew, by loud cawing, whether the flock ought to go on flying or land somewhere. Anyone out for a walk in the country during the winter can observe this process. In every flock of crows that is starting a flight, flying, or landing, two cries of entirely different timbre can be heard. One is a high, hard clicking that also has an agitating effect upon human beings. The other is a sulky-sounding caw. The first expresses the desire to fly, the second weariness, hunger, desire to land.

In the incessant alternations of these announcements of mood and avian 'opinions', however, a simple 51 percent majority is not sufficient. It is astonishing to observe that although the large mass of crows are eager to rest, they can be repeatedly inspired to fly on a little longer by a fairly small number of active members of the flock. This continuous vote in the air can go on for hours sometimes, until transmission of mood and actual fatigue produce the necessary vote of three quarters in favor of landing.

Professor Hubert Fringe[18] and his wife Mabel of Pennsylvania State University demonstrated that majority decisions were involved by trying to influence the voting with a loudspeaker. The zoologists tried to induce flocks of crows flying over an open field to land by loudly playing the 'rest call' on a tape recorder. The ruse was hardly ever successful. Only once, when very nearly the requisite majority in the flock was already pleading for a landing, were they able to swing the balance. The cawing from the loudspeaker added a final additional vote, and so the birds landed in the vicinity of the recorder and the tent where the scientists were hidden.

One might imagine that such democratic decisions are practiced only by a few of the 'intellectuals' in the animal kingdom. On the contrary, and significantly, this phenomenon is by no means restricted to intelligent species. The reason is that an individual can survive only if he is able to live under decent conditions. That is just as true for primitive as for more advanced forms.

Insect societies in particular are all too often pictured as examples of the starkest totalitarian dictatorships. But this view is based on insufficient information. To this day no scientist has been able to discover a dictator in any insect community. The queen is rather an egg-laying 'machine', although she probably regulates some of the processes in her society by emanating various odors. She is certainly no dictator. On the contrary, modern studies have revealed types of democratic decision-making among bees, ants, and termites. There are indications that an ideal kind of socialized feeling prevails in these societies.

Virtual parliamentary debates occur, for example, in swarming bees. Here, the choice of a new dwelling is a life-and-death decision for all. If the swarm left such a decision to a single member whom it obeyed blindly, bees would long ago have disappeared from the world. Instead, the following procedure takes place: as soon as the swarm has moved out of the old hive and formed the familiar cluster on the branch of a tree, some forty of the approximately twenty thousand bees in the swarm fly off to search for a new home. These bees investigate every cavity, every box, garbage pail, and cardboard carton, every hollow tree, and crevice in a wall.

When one of the scouts has found a place she thinks suitable, she returns to the waiting cluster and by dancing on the surface of the cluster informs the other returned scouts of the direction and distance of the new domicile. As Professor Martin Lindauer[19] of the Zoological Institute of Frankfurt am Main has determined, the forty-odd scouts are far from unanimous at the beginning. Virtually every individual makes a different proposal. But those who are not especially enthusiastic about their own discoveries fly out to inspect the offers of the rival scouts. The tiny creatures then make their own decision, whether to accept the proposal or to look for further prospects.

Thus parties form. They may be recognized by their similar instructions in dance language for direction and distance. At first there are many small factions. As the number of factions gradually lessens, the adherents of each increase. In vehement, factual discussion they debate the question of which site is to 'house' the entire swarm.

Sometimes these debates continue for days, with repeated inspections of the various prospective sites. It may happen that the home for which the largest party pleaded at one time is not accepted in the end because a smaller group has managed to persuade the majority of the advantages of its proposal. There is no obligatory party discipline. The great move does not begin until the forty scouts have reached full unanimity.

Ants strike up mutual 'agreements' on the tempo of work. Dr. S. G. Chen[20] selected two ants of the same nest with opposite labor ethics. When each was kept by itself, one moved extremely slowly in bringing up morsels of earth for the nest building, whereas the second worked very rapidly. But when they were placed together, the tempo of the fast worker immediately slowed, while the lazy one felt obliged to speed up somewhat.

There are even ants for whom the term 'queen' is far too absolutistic, since their ruler is chosen out of the horde of common workers. These are the feared South American army ants of the genus *Eciton*. A single host can comprise up to twenty-two million ants. In their predatory expeditions they devour all creatures in their path who do not flee in time.

Every so often such a nomad army of ants divides into two marching columns of equal size. A few days before the division takes place, several still immature queens—called princesses—hatch out in the bivouac. But only one of them can become queen. In a similar situation among bees the candidates sting one another to death until only one is left. But among the army ants the prospective queens compete during the course of the expedition. Dr. T. C. Schneirla[21] of the American Museum of Natural History in New York observed how each princess contrived to interest a constantly changing band of followers in her cause. Gradually an order of rank is established among the candidates for the throne. What qualities on the part of the princess decide the issue is not yet known. Evidently certain types of behavior in critical situations during the expedition, and certain attraction and signal scents emitted by the princess, play a crucial part.

Once the decision has been made, approximately half of the army of twenty-two millions leaves the old queen, forms a new, independent ant nation, and marches off in the opposite direction with its newly chosen young queen. Soldiers force the superfluous princesses to the rear of the great army, where they are abandoned by all. Exposed like some ill-omened royal progeny, they soon die in solitude.

Here we must stress once more that democracy in ant society must not be understood in the human sense. It does not spring from ethical considerations. Nor is it based on personal liking, as is the case among apes and prairie dogs. Rather, it is an innate instinctual mechanism that functions more or less 'automatically'. Democratic behavior is programmed, so to speak, into these robot-like creatures. But the end result comes to the same thing: the chance to choose the best among many is of considerable benefit to the community. That is all we wish to emphasize in this context.

The same principle holds true for another remarkable phenomenon in insect societies: socialized feeding, which in its consistency goes far beyond the most idealistic human notions.

In the interior of these heavily populated states the ants' supply of food is allocated down to the last ant, with a 'justice' inconceivable to us, by a continual process of begging for food and giving it away. Professor Karl Gösswald[22] writes: 'If a hungry ant meets a companion whose crop is filled, it stops the other with lively motions of its antennae, rapidly vibrating its feelers against the other's head, caressing the other's cheeks with its forelegs, and licking around the other's mouth. This behavior means in ant language: "Give me something to eat." An ant subject to this begging lays its own feelers back, opens its jaws, extends its tongue, and produces a drop of fluid which is greedily licked up by the hungry ant. Every ant that is fed in this way will in turn pass on the greater part of the food to other ants.'

The hungrier an ant is, the less often will it voluntarily offer a share of food from its socialistic stomach to other members of the nest and the more often it begs from all who pass by it. But if a very hungry, begging ant encounters a fellow who begs even harder than itself, it immediately gives up part of its store, no matter how little it has. Thus the state of nourishment within a nest is always equal. Idlers receive exactly as much to the milligram as the most diligent ants. Special services to the state, such as tracking down booty or fending off enemies, are not rewarded by the least additional ration. Here is communism without those little extras for functionaries and party bosses!

This whole procedure takes on additional importance in the light of a discovery made by Dr. Rolf Lang[23] of the University of Freiburg im

Breisgau. For he has found that socialistic feeding does not, despite its absolute 'justice', operate blindly and without regard for individuals.

Red ants gather two types of food: animal proteins in the form of insects and honeydew which they milk from aphids. But distinctions are made: both kinds of food are not distributed evenly to all members of the nest. The workers in the interior receive almost exclusively protein, since they need growth substances for the offspring in the nurseries. The food gatherers on outside duty, who chiefly need energy for their muscles, receive the overwhelming portion of the aphid milk. Within each of these two groups, however, each ant receives a precisely equal portion of the 'social product'.

Only one sad exception to this rule has been discovered so far.[24] There are parasites who have insinuated themselves into the ant society. These are small beetles whose physical form bears no resemblance to that of the ants, but who have developed a perfect system of identifying themselves as ants by giving off the right scents and making the proper begging gestures. They lurk around in the dark corridors, do nothing productive, and beg, beg, beg. Since they appear to be the hungriest of all, they are readily fed by every passing ant, receive many times what a rightful member of the nest deserves, and grow fat on doing nothing —even in times of famine when all the ants are suffering from hunger. Shall we say that they provide a perfect example of the misuse of democratic institutions?

A rove beetle (right) convinces an ant by emitting scents and by using the tapping language of the antennae that he is also an 'ant' and wants to be fed.

Rescuers, Midwives, and Friends

When men dig underground shafts while mining coal or ore, disasters happen every so often. Tunnels collapse, burying workers or cutting them off from the outside world. Similar catastrophes can easily befall the miners of the insect world, the ants.

After heavy rainstorms, especially in tropical regions, tunnels frequently collapse in the underground labyrinths of leaf-cutting ants. Such troubles are so common that the leaf-cutters might long ago have become extinct if the small creatures did not provide mutual aid to one another in emergencies. Buried ants produce SOS signals. Rescue squads at once set off in search and dig them out.

Aid to comrades in danger! Who would have expected anything of the kind in the supposedly soulless, robotlike ant societies? In this instinctual behavior of the small insects, nature has created something which in its ultimate effects comes to the same thing as humanitarianism in men. For such helpfulness can be a highly utilitarian trait in the struggle for existence. It may even be of crucial importance and therefore a quality which has been 'bred' in the course of evolution.

The SOS calls are uncommonly loud, considering the size of the insects. Dr. Hubert Markl,[25] the Frankfurt zoologist, measured these calls in the course of a study undertaken on the West Indian island of Trinidad. At a distance of a centimeter from the buried ant the distress call sounded as loud as an ordinary typewriter.

Man cannot hear it without instruments, however, for the ants broadcast on the ultrasonic level between twenty and 100 kilohertz. They rub rough surfaces of their abdominal segments together to produce this inaudible—to us—sound. The rescuers, workers, and soldiers, do not receive these distress calls through the air (which would be pointless in the case of the buried victims of the disaster), but as vibrations in the ground. Their 'ears' consist of vibration-sensitive cells on their feet, which enable them to trace the source directionally. Thus the search squad can determine precisely the position of the buried ants.

The defense mechanisms of leaf-cutting ants are even more elaborate. Every year in March these ants on Trinidad are exposed to a special danger. In the late afternoon hours tiny flies attack the worker ants like dive bombers, flashing down on them while the ants are cutting up a leaf or carrying their 'sails', consisting of bits of leaf, in long processions from a tree to the nest. Since the ants are clinging to their harvest with their mandibles, they are completely defenseless.

The flies richly deserve their name of 'ant beheaders'. They make their lightning assault in order to attach an egg to the victim's neck. Shortly[26] afterward the fly larva hatches out and bores into the ant's

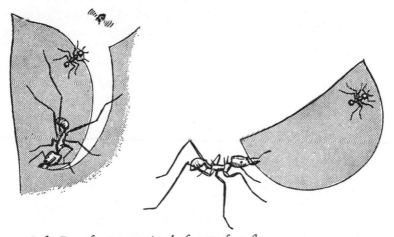

Left: Dwarf ant protecting leaf-cutter from fly.
Right: Dwarf ant, serving as antiaircraft defense, rides home on leaf.

head, gradually eating it hollow and finally biting off the hollowed-out head. These parasites behead their host in order to pupate in the capsule which is all that remains of the ant head.

The ant society has developed a regular antiaircraft squadron to deal with these deadly onslaughts, as Dr. Irenäus Eibl-Eibesfeldt and his wife[27] have shown in a fine series of pictures. Air defense is provided by a group of ant dwarfs who attain only a fourth the size of the average worker ant. As a rule these dwarfs work as 'farmers' in the underground fungus gardens. But shortly before the time of day at which air attacks may be expected, they leave the nest and accompany the parade of their larger fellows to the site of the harvest, say a grapefruit tree.

As soon as the flies attack, several dwarfs form a protective wall around a leaf-cutting worker. They rear up and snap at the enemy with their mandibles. In this way they successfully fend off many attackers. When a worker ant has cut out a piece of leaf, one of the dwarfs climbs to the upper edge of this piece and hitches a ride back to the nest—acting as a kind of portable antiaircraft gun.

Helpfulness is found not only among ants, but also among elephants. Many odd stories are purveyed concerning the huge pachyderms, and it is often hard to know whether big-game hunters are not amusing themselves by telling tall tales. Fortunately, however, we also possess some authentic recent reports.

On the slopes of Mount Elgon in Kenya an elephant herd of some thirty bulls, cows, and young elephants was threatening to increase beyond the capacity of its range. Toward the end of 1964 the game warden, H. Winter,[28] therefore had to shoot three of the animals. When the fatal shots were fired, commotion broke out among the survivors. They wheeled about and trumpeted and screeched terrifyingly. Stones, sticks, and clumps of soil flew through the air. Then they tried to raise their dead fellows from the ground.

Warden Winter writes: 'The bewildered and enraged beasts pushed and butted ponderously at the inert forms and, by entwining their trunks with those of the dead animals, tried to lift the carcasses. They also pawed at the dead with their forefeet in a frenzied effort to move them. I watched this performance for a little more than a quarter of an hour, throughout which time the elephants persisted in their attempts at moving their stricken comrades, showing no other interest or fear.

'Presently, a large cow advanced to where one of the elephants lay, and knelt beside it, placing her tusks under its belly. She then tried to stand, her body tensed with the tremendous effort. Suddenly there was a loud crack and her right tusk snapped off at the root, describing an arc through the air and landing about thirty feet away.'

In order to help the bull, which might have been her husband, the elephant had sacrificed a tusk.

Shortly after this incident the herd left the site, roaring fearfully, and smashing all the trees in its path. Suddenly the giants returned, once more trumpeting and screeching and trying again to raise their dead fellows. Three times they repeated these efforts. At last the leader of the herd stood majestically still and uttered several signal cries, as though to rouse the dead elephants. When this, too, had no effect, he trumpeted loudly, and the herd at last moved heavily away.

In 1966 game wardens[29] of the Addo National Park in South Africa reported a particularly stirring case.

An old cow elephant who went by the name of Oma had for some time been suffering from a tumor two feet in diameter. She became so emaciated that she could scarcely stand on her feet. Her son Lanky supplied her with fresh green fodder every day, and led her to the water hole by pushing her from behind or supporting her on the side toward which she tended to topple. Finally, when Lanky could no longer manage her alone, the entire herd accompanied the feeble old cow to the waterhole.

A game warden decided to attempt to operate on the sick cow. He stunned the animal with an arrow dipped in a narcotic drug. Lanky helped his tottering mother to lie down in the shelter of a thicket.

What happened next has been described by the eyewitness: 'Soon afterwards the head bull appeared and forced his way into the thicket. A few seconds later he reappeared with his tusks bloody. Supposing the cow on the verge of dying, he had delivered the *coup de grâce*. With a loud cry he then summoned the entire herd. All the elephants formed a circle around the dead cow. After another cry from the leader, the herd left the scene.'

In view of what we know about elephants, the expression *coup de grâce* is not an invalid example of anthropomorphic language. We have just pointed out that the behavior of the leaf-cutting ants is, in its effect, equivalent to human moral behavior. In the case of intelligent elephants we need not speak of 'equivalence'. These mighty beasts are so closely united within a herd by ties of personal friendship and a sense of belonging that altruism, helpfulness, and loyalty unquestionably exist among them.

Baboons, too, manifest such traits. Norman Carr,[30] director of the Kafue National Park in Zambia, relates the following instance:

I have watched a leopard scatter a troop of baboons, sending them scurrying up trees—all except a youngster who was cut off from retreat. I then saw a female baboon, at the peril of her life, descend the tree to grab the terrified squealing baby and rush back up the tree with it while the rest of the troop, from the safety of their lofty perches, were swearing and making rude gestures at the rather bewildered leopard who had mistimed its attack. If this had been a mother protecting her young I could have understood the action. But I had been watching these baboons for some time before the assault, and I am sure that the gallant baboon who came to the rescue of the youngster was not its mother.

Many pages might be filled with examples of this kind. We have already seen how chimpanzees remove a wounded member of the band (page 12) and free their fellows from traps (page 12). On pages 25 and 35 we have spoken of the rescue actions among dolphins.

In this connection Dr. John C. Lilly, the American delphinologist, reports an especially remarkable example. (*See* Chapter Two, note 1). A bottlenose dolphin had contracted a cramp in the cold water of a freshly filled pool and could swim only in a twisted position. At first the dolphin's behavior followed a familiar line: it emitted SOS signals, and instantly two friends came to its aid ready to hold it above water in their usual manner. But this was no ordinary injury. The dolphin could breathe perfectly well. But it could not swim straight. Once again it uttered several whistles. The two helpers answered, then promptly thrust their bodies against a spot just in front of the patient's tail fin. Experiments have shown that pressure on this spot produces a muscular reflex that extends the body. So it was in this case. The episode proved that the dolphins' rescue maneuver was not, or not only, an instinctual process. Along with it went a conscious intention to help, and apparently something that may be regarded as medical skill.

Dr. Erna Mohr[31] reports amazing things concerning another marine mammal, the walrus. 'This unqualified mutual aid that walruses offer one another, their tireless support of one another in all situations, is one of the most remarkable traits of these intelligent animals. Wounded walruses are not abandoned. Fellows that have been shot on land are dragged into the water if possible. A strong member of the family will swim underneath an animal incapable of swimming and support the weight of its neck so that the wounded walrus can keep its head above water. Mothers tuck their cubs under their flippers and carry them off to safe waters.'

One especially amazing item on the roster of animal altruism is midwifery. 'Gynecologically skilled aunts' are as much an institution among the spiny mice as midwives in human society. The spiny mouse, an African relative of our domestic mouse, has very difficult parturitions, since its young are unusually large and, moreover, are born by breech presentation. Without the aid of experienced females, infant mortality would probably be so high among these animals as to threaten the survival of the species.

Dr. Fritz Dieterlen[32] of the University of Freiburg im Breisgau has recorded the following details. The initial labor pains last an unusually long time. Unlike many other species, however, the prospective mother does not isolate herself from the rest of the clan. Rather, she awaits the birth in the midst of the others, who usually keep close watch on the procedure as parturition begins.

Here is Dr. Dieterlen's record dated May 29, 1961:

... 4:17 o'clock the curve of the baby's hindquarters appears. The mother licks it. Then the hindquarters recede again. 4:18: continual pressing. 4:19: the hindquarters and hind legs half appear, then emerge fully. While they are dangling, the surrounding membranes tear and are licked away by the first midwife. 4:21: the baby slides farther out. 4:22: after extremely vigorous pressing the body has slid out toward the rear. The midwife quickly licks it clean. But the head is still inside. The mother presses and presses, but makes no further progress. Nor can the others of the litter begin their passage.

4:29: a sudden jerk, and the baby is ejected. It brings the afterbirth with it. This is eaten by the first midwife and another who has come up. Then all clean the newborn mouse.

In many cases the midwives also gnaw off the umbilical cord. The zoologist comments on the value of this gynecological assistance: 'Even in the case of difficult births healthy females, if they pause between exerting pressure, are able to lick the membranes away from the emerging snout, so that the baby can breathe. But they cannot simultaneously press. If midwives are present these pick away the membranes while the parturient presses. This saves time.'

Close observation of eighty-six births showed that all the 'midwives' had already had at least one litter. Females which have not yet given birth remain totally indifferent to the whole process. This indicates that the assistants do not eat the sacs and afterbirths merely out of hunger; they are acting out of experience and actually want to help.

Incidentally, if the mother mouse dies soon, or later, the midwives adopt the litter. The young are thus provided for.

At the birth of dolphins[33] also, two 'midwives' must be present and hold the offspring up to the surface of the water for it to breathe. Llamas and elephants[34] likewise take a keen interest in every 'happy event' in their herds. Among African elephants,[35] cows are sometimes said to help the parturient in removing the amniotic membrane and getting the baby to its feet. Among South American marmosets it is the male who provides aid.[36] The male obstetrician catches the infant with both arms and affectionately cleans it.

Sometimes an entire herd offers a highly useful form of assistance at a confinement, though it could scarcely be called obstetrical. Maurice Fiever, who in collaboration with the author of this book has filmed a movie of African animals in the wild, once witnessed the following scene.

A calf had just been born in a herd of gnus. From the moment of birth to the moment the calf can stand up on its long, sticklike legs and walk, thirteen minutes pass. Those are the most dangerous minutes in the baby gnu's life. For when gnus are giving birth, all sorts of predators—hyenas, jackals, and lions—prowl around the vicinity of the herd, looking for easy prey.

Within a few minutes a hyena scented the favorable opportunity and approached. Whereupon a fascinating reaction ensued. Some of the adult members of the herd observed what was afoot and instantly formed a broad front to protect the calf, which was still thrashing about on the ground, from the hyena. Gnu beside gnu, they presented to the foe a formidable row of sharp horns. When the predator saw this, he resignedly sauntered away.[37]

Immediately afterward, a lioness attempted to approach the newborn calf. Normally, the animals of the herd would have galloped off in wild flight at the appearance of the big cat. But now, with a calf to protect, they promptly and heroically closed up in a defensive front again. Though the lioness crept closer and closer, the gnus did not retreat. That courageous defiance made its impression upon the 'queen of beasts'. She did not dare to spring into that horned phalanx, and took herself off.

'In union there is strength.' This principle holds equally true for the animal kingdom. Wild geese, when they are numerous, will advance in a close battle line against a fox. In the face of this determined spirit, even an animal so passionately fond of geese—between his jaws—will have no choice but to retreat.

Toward the end of the eighteenth century the springboks[38] in South Africa were still so numerous that they loped through the plains in herds of up to fifty thousand animals. According to reliable accounts, when inexperienced young lions attempted to attack them they are said to have encircled the predator and forced him to march along in their midst—until he collapsed from hunger and exhaustion and was trampled to death.

'The reaction of social attack against the wolf is still so ingrained in domestic cattle and pigs,' writes Professor Konrad Lorenz,[39] 'that one can sometimes land oneself in danger by going through a field of cows with a nervous dog which, instead of barking at them or at least fleeing independently, seeks refuge between the legs of its owner. Once when I was out with my bitch Stasi, I was obliged to jump into a lake and swim for safety when a herd of young cattle half encircled us and advanced threateningly. . . .'

Professor Lorenz also relates that his brother, with his Scottish terrier tucked under his arm, once had to scurry for safety up a tree when a herd of half-wild Hungarian pigs, 'with bared tusks and unmistakable intentions,' advanced menacingly from all sides upon the man and the dog.

In the light of these accounts a 1967 report from New Delhi does not sound like a tall tale. Near the town of Surat in western India a fifteen-year-old shepherd was attacked by a tiger. The sheep promptly gathered in a dense formation, ran at the surprised feline, and trampled it to death. The terrified boy came out of the affair uninjured.

On the same principle, many starlings are death to the hawk, as the ornithologist E. Gersdorf[40] has observed. He has often seen the misadventures of a hawk who has had the bad judgment to attack some starlings when the whole flock, consisting of some thousand birds, are in the same field. The scattered birds promptly form an assault squadron which 'overwhelm the sparrow hawk as it is circling the field and fly on with undiminished speed. After perhaps only a second the hawk loses his leading edge. The squadron then wheels around and attacks it anew, forcing it to land.'

Once a sparrow hawk that had been forced down flew headlong into a bush in its panic and had some difficulty afterward in getting out. Another sparrow hawk pursued by starlings was driven into a lake. With awkward swimming movements it managed to reach a zone of reeds. But here it could not properly take off into the air again. In searching through the reeds the ornithologist found three dead hawks.

Paradoxical though these examples of attacks by the weak upon the strong may sound, they can be plausibly explained. Danger intensifies the sense of belonging and identity. Being part of a mass restores courage to the individual—anonymous, mass courage, to be sure, but in such a state he is ready to join with others in a daring action that works to the benefit of all.

The earlier cited cases of animal helpfulness are more difficult to interpret. Why, we must ask, do elephants, dolphins, walruses, chimpanzees, and other animals help injured and sick members of their species? Heretofore it has been a generally accepted tenet of post-Darwinian thought that precisely the opposite behavior is typical: killing of wounded, sick, and in general any abnormal member of the species, extinction of 'worthless life', hostility toward all members of the species who do not conform to the norm. And in fact a chicken can arouse the hostility of all other chickens if the quill of a single feather in its head is bent and sticks from its plumage in the wrong direction.

But for the very reason that many animals kill dying members of their species, aid given by an animal to a fellow in need constitutes a tremendous achievement and a significant step in the direction of human behavior.

The subject is again illuminated by the conduct of crows. In the fall of 1964 Nuremberg newspapers reported that crows in the city park had seemingly become 'rabid'. For several days they had been diving furiously at elderly ladies walking alone and striking at the unfortunate women with their beaks—in scenes reminiscent of Alfred Hitchcock's film *The Birds*.

What was going on? Behavioral scientists have learned that when members of a crow flock miss one of their companions, they search the entire area for it for several days. Every predator, including man, is suspect during this period of search. And should an innocent stroller be swinging something black, say a black handbag, the crows think they are seeing their missing fellow in the grip of an enemy. They promptly try to liberate it.

Daws behave in exactly the same way. They also are impelled to come to the defense of one of their species, and are easily misled by a black handbag.

The more intelligent ravens[41] react quite differently. They likewise will spend a long time searching for a missing fellow. But they attack only when a raven is actually in danger *and* when they recognize this raven as a personal friend. If, for example, a bird-fancier catches a strange raven and holds the struggling bird in his hand, his own raven will approach to help the man, whom he knows, 'kill' the stranger. He will side with the other bird only if he recognizes him as a friend.

Among ravens, then, helpfulness has been freed from the rigid fetters of instinct (as we have seen it among leaf-cutting ants, crows, and daws) and has evolved into a semimoral intellectual achievement. The decisive element has become personal friendship.

But—is there such a thing as personal friendship between animals? Let us listen to the comment of the zoologist Dr. Otto von Frisch:[42]

When my fawn Babette was three weeks old, a remarkable incident occurred. We were dining on the terrace, and Babette was grazing on the lawn beside us. Suddenly all the birds in the garden fluttered in alarm. My tawny owl, Kiwi, whom I had never seen leave his tree by day, plummeted down from his hiding place, landed on the bare branch of an apple tree, and eagerly eyed the tiny fawn.

My first thought was that Kiwi had evil intentions. But after about five minutes the owl hopped one level lower, from which a brief gliding flight brought him into the tall grass right in front of Babette. Now the fawn at last noticed. Cautiously, setting one leg in front of the other, she approached the owl, extending her head toward him, and licked him once across his face. Then she turned, and with several graceful leaps vanished behind the house. Kiwi, startled, sat there with open beak and an expression that can only be compared to that of a shy young man who has just received a first kiss from the girl he has secretly adored.

I had witnessed, as I soon realized, the beginning of a friendship between two animals in which man had not played the part of mediator.

It had arisen out of the free and independent decision of both participants.

No human being can ever say why the owl had set his heart on the fawn, and vice versa. What could have prompted such utterly different animals to form an inseparable couple henceforth and to go everywhere and do everything together? We do not know. But it is precisely the sign of genuine friendship that it is not based on material interests or upon the fact that both partners have grown accustomed to one another by being forced to live together. Here we have proof that animals, too, can be seized by a kind of 'fondness' for one another—just as human beings can be. And what applies to fawn and owl certainly applies at least to an equal degree to members of the same species. Such friendships are found among ravens, elephants, dolphins, wolves, and many other animals.

Those who think they can buy an animal's friendship by feeding it need to be told bluntly: the animal regards you much as a cow regards her pasture. True affection has nothing to do with feeding—no more so for animals than for people, whose love cannot be won by gifts, wage increases, or rounds of drinks.

When, however, a genuine friendship between very different animals arises, there is also great danger of misunderstandings. One example of that occurred in 1963 at the Hagenbeck Zoo in Hamburg.

Here, too, it was a case of affection at first sight. One day a stray cat sat down in front of the ape cage. The gibbon, who, sharing his cage with a female, was not at all lonely, promptly swung along his horizontal bar, reached through the bars, and stroked the cat. He caressed her in a convincingly emotional manner. Henceforth, the cat visited her gibbon every day for weeks. But then the friendship came to a regrettable end. Thanks to her strength and litheness, the cat managed to force her way between the bars to the object of her affection. But as soon as she entered the cage, the ape threw his arms lovingly around his feline friend and refused to let her go, no matter how the cat hissed, meowed, and clawed. It was all of two days before the cat, by this time very hungry, succeeded in liberating herself from that loving embrace. No wonder the friendship ended the moment she escaped.

Misunderstandings likewise account for dogs and cats behaving 'like cat and dog'. It is simply not true that hereditary enmity exists between these two species. Aside from those rare people who take delight in inciting their dogs to chase cats, the reasons for the animals' hostility are to be found in their differing behavior patterns.

If a dog wants to play, he lifts his forepaw and wags his tail. But in cat language both these gestures mean the very opposite: 'Beat it or I'll scratch you!' Therefore the cat takes care not to play with the dog.

The same misunderstanding occurs when the cat indicates her readiness to play by purring. The dog interprets the sound as growling, therefore as an intention to frighten him away, and so no friendly contact ensues.

When, to compound the difficulty, an irritated cat defensively raises her paw, the dog thinks she wants to play with him and prances up to her—only to have her claws rake his face. What happens afterward, and in all such future encounters, is only too well known.

Fortunately, these misunderstandings can be overcome. This fact is one of the discoveries of animal behavioral research which has its applications to human social life. If dog and cat grow up together from infancy and have some degree of social intelligence, they learn to understand each other's gestures. Each recognizes the language and the crotchets of the other and knows how to adapt to them. That is the whole secret of the many outlandish friendships between dogs and cats, foxes and geese, crows and guinea pigs, martens and chickens, tigers and horses in the circus, and so on.

Working along these lines, Professor R. Menzel[43] at Kiryat Haim, the Israeli institute for the study of dogs, established a friendship among three animals, a dog, a cat, and a rooster. Every morning the rooster enthusiastically greeted the dog and was licked by it. Then he scratched the ground for feed and summoned the cat by crowing as if she were a hen. The cat obeyed the summons. But since she was not especially enticed by corn and worms, she merely rubbed amicably against his feathers.

Only one animal was missing to transform this idyll into that prototype of animal friendship, the Musicians of Bremen: the donkey.

4. Population Policy in Animal Societies

Restrictions on Marriage

At Cape St. Mary on the precipitous coast of Newfoundland the air is
filled with the screeching and flutter of thousands of seabirds. Innumerable nests of gannets, birds as big as geese, are crowded close
together on the rocky walls enveloped by the surf in perpetual clouds
of spray. It might seem the center of some dire population explosion.
But appearances are deceptive.

The Scottish zoologist V. C. Wynne-Edwards,[1, 2, 3] Professor at the
University of Aberdeen, has studied these vast colonies of breeding
gannets and has noted something quite strange. Only the birds who on
arrival in the spring have been able to win a nesting place on a cliff that
juts particularly far out over the water, are allowed to celebrate their
wedding, lay eggs, and raise offspring. All the others, those who have
come too late and those who are displaced by stronger birds, are
banished to a neighboring cliff. Here, too, males and females are together and it seems as if there is nothing to prevent their building nests
and having young. But they do not do so. It is as if an invisible borderline surrounded the traditional site where the birds have bred for
millennia, and as if all the birds outside that line were under the spell of
a rigid sexual taboo.

Here we have a truly fantastic discovery. For many people, any type
of birth control is branded as an unnatural act. But scientists have by

now demonstrated that almost all animal species prevent overpopulation by instinctive behavior. The practices of the animals range from simple continence beyond a restricted number, so that any population surplus is kept from breeding, to birth-control drugs, and even to cannibalism.

Of their own accord, so to speak, gannets limit their number so that under normal conditions all members of the species, including the involuntary virgins and bachelors, find enough fish in the surrounding sea. What is more, all losses in the breeding colony can promptly be balanced out, for as soon as a nesting site becomes available a pair moves in from the reserve of single birds on the other cliffs. A similar 'social equilibrium' also prevails among penguins, shearwaters, guillemots, and oyster-catchers, and is even found among the common songbirds of our woods. But the method of control is not easy to observe.

The zoologists M. M. Hensley and J. B. Cope[4] became aware of these processes for the first time when they attempted to keep a smallish zone of woods completely free of birds. As soon as a pair indicated by its singing that it had occupied breeding territory in the zone, it was shot down. But next day the territory was occupied anew. Apparently there had all along been a numerically large group of birds, condemned to silence and disqualified for mating, present in the dense undergrowth of the wooded zone. From this reserve new pairs could move into the breeding territory as soon as there was room.

Gray seals practice a form of restriction somewhat similar to that of the seabirds. About four thousand of these seals come together annually on the Farne Islands off the North Sea coast of England. Here the seals crowd together in inconceivable density on the islands they have traditionally occupied. At first glance the entire surface of these islands seems to be covered completely with seal flesh. Many newborn calves are carelessly crushed by the adults; many others lose their mothers in the tangle of bodies and starve to death. In short, what we have here are all those fatal symptoms of degeneration associated with overpopulation. And yet there is no need for the gray seals to crowd together under conditions of such intolerable density. As Dr. J. C. Coulson and Grace Hickling[5] of Durham University have pointed out, there are five other uninhabited islands in the immediate vicinity which would serve as ideal seal breeding grounds. But the animals pay no attention to these other islands.

In order to prevent the exhaustion of their fishing grounds, the gray seals seem deliberately to have produced a situation in which over-

Gannets may mate and raise offspring only on the center cliff (shown white) on Cape St. Mary. Hundreds of the birds perch on the white spot of the cliff to the left. But here a strict sexual taboo prevails.

population exacts its accompanying toll. But they do this only while raising their offspring. The astonishing fact is that the animals institute their population controls before the food supply becomes so scarce that the weak face starvation. It is not present hunger that limits their numbers, but the threat of hunger in the near future.

What would happen if animals did not have this 'foresight' rooted in the structure of their instincts has already been revealed to us in the case of an animal whose greed has blurred his sense of the harmonies of nature: man. Denuded hillsides, tillable land reduced to desert, animals slaughtered and exterminated by the millions, mark where man has passed. Professor Wynne-Edwards concludes that if all animals overhunted, overgrazed, and exploited their sources of food as do man and locusts, all living things would long ago have vanished from the earth.

In saying this the scientist is deliberately taking issue with the view of Charles Darwin who recognized, as the counterpoises to boundless fertility, only externally operating forces: death by hunger, predators, storms, and disease. Granted, these apocalyptic horsemen take their toll. But to Professor Wynne-Edwards' mind, they are not the ultimately decisive regulators of population density.

As far as we can judge today, the hypothesis of self-regulation casts a wholly different light on our view of the dynamics of nature. The time has come finally to take leave of those complicated old theories[6] which a wit in 1880 parodied as follows:

In order to strengthen the British fleet, many men went to sea. To feed them, the Admiralty needed a great deal of meat from cattle that were raised on clover. Expanded production of clover had a favorable effect on the distribution of bumblebees, which collected honey in their underground nests. But the mice fed on the honey and increased enormously—to the delight of the cats. Therefore many old maids were able to keep cats and thus console themselves for the fact that so many men had gone into the Navy.

Of course this was only a joke. But still it was believed that, for example, an increase in the bumblebee population of only ten insects would produce changes—minimal changes, to be sure—in other populations and that these in turn would affect others.

Running completely counter to this line of thought is the hypothesis advanced by two Australians, H. G. Andrewartha and L. C. Birch,[7] in 1954. In their own country they had too often seen how overwhelming reproduction by a wide variety of species flooded vast areas until there would be billions of animals perishing of thirst in a given region. The whole population would die out until one day there came a new influx from a distance. The Australians argued: Balance of nature is a myth. Even minor climatic variations can cause animal populations to explode or die out.

Such a hypothesis may gain credibility under the extreme conditions of their continent. But closer studies, based upon the classical example of the rabbits,[8] have shown that the hypothesis does not fit the facts. Granted, shortly after their importation into Australia, the rabbits launched upon a terrifying reproduction boom. The twenty-four rabbits that Mr. Thomas Austin set free on the new continent in 1859 became twenty-two millions within six years! And that was only the beginning of the rabbit explosion! By now the furry creatures have conquered the continent to its remotest nook; but it is significant that the rabbits are no longer breeding 'like rabbits'. For the fertility of these animals is dependent on the weather. In times of extreme drought the males do not approach the females. They simply practice continence. For in such conditions the offspring would not survive. If a pregnant rabbit experiences extremely hot, dry days, it is subject to a kind of stress, and miscarries. Immediately after the first rainfall the rabbits' fertility returns to them in full measure.

This would appear to confirm Wynne-Edwards' theory that animals themselves regulate their numbers in response to conditions.

There is scarcely any parallel in the animal world to the lifelong near-starvation of millions of human beings in underdeveloped countries—side by side with 'holy' cattle and other domesticated animals. Animals in the wild experience hunger only temporarily, in periods of distress during unusual cold or drought. The only exceptions are those species in which the regulatory mechanism is too weak or has become perverted: locusts, some butterflies, noxious insects, and lemmings.

All this is a far cry from the concept of life as a struggle for existence, an eternal war of all creatures against all others. Yet evidence mounts that most animals do not reproduce like lemmings, locusts, and human beings because they limit their own capacity for reproduction. The fact is borne out in every cage and aquarium. Whether it is fruit flies, meal beetles, water fleas, guppies, rabbits, rats, or mice that are being kept, and offered an inexhaustible supply of food and given optimum conditions, the same phenomenon will be observed. At first their numbers rise sharply. But soon the rate of growth slows down, and finally the population density remains constant or else even diminishes again 'of its own accord'.

What prevents these creatures from giving free rein to their fertility? Why do even rats not multiply 'like rats'?

Professor John E. Calhoun[9] placed twenty male and twenty female rats in an enclosure of 1,000 square meters, which he made a veritable paradise for them. After twenty-seven months the rodents should have multiplied to about 5,000. The space and the food supply would have permitted such reproduction. Instead, at the end of that time only about 150 adult rats were living in the enclosure—and the population remained fairly static thereafter.

What is the explanation? As soon as the population density passed a certain limit, the otherwise surprisingly good manners and morals of the animals disintegrated. Males raped the females. The females stopped building nests, but brought forth their young on the hard ground, neglected them at the first provocation, and ceased to pay attention to the screeching babies. Finally the young were eaten by roaming males. The consequence was: infant mortality of 96 percent, maternal mortality of more than 50 percent, and premature death of many of the males from stress, total exhaustion, or savage fighting. And all this took place although there was still feed enough and room enough for all! Such cannibalism likewise appears in its ugliest form among closely packed animals in stables, cages and aquariums.

Professor Otto Koenig,[10] the Viennese behavioral scientist, witnessed similar excesses when he attempted to investigate the question of what would become of a social order living in eternal prosperity. In an enclosure at the Wilhelminenberg Biological Station he established a heaven on earth for his colony of buff-backed herons (*Ardeola ibis*), providing it moreover with a constant surplus of food. But it soon turned into a hell.

The social order and family life of these birds fell into total confusion. While the sexual activity of the crowded colony increased to a grotesque degree, the numbers of offspring rapidly diminished. The parents, which live in strict monogamy in the wild, seemed to have nothing in mind but adultery, triangular and quadrangular relationships, polygamy, rape and incest, quarrels with their neighbors and even within the families. Filthy and bleeding from fights, they trampled the eggs in the nests and let the chicks die.

The young that survived nevertheless did not even learn to provide for themselves. The only thing that linked them to their three or four 'parents' was incessant begging for food. Even when they were grown, they pursued their elders to the eternally filled feed trough and insistently begged. Probably only to be left in peace, the parents then gave something to the infantile adults. When the latter in turn had offspring, they were incapable of caring for them. The grandparents had to feed both children and grandchildren.

Professor Koenig fears that if automation does away with work on a vast scale and confers boundless prosperity upon the human race, there would be a drift toward a similar degeneracy. Would man's celebrated reason check any such development?

Further studies of animals open new perspectives. Here is one startling discovery in this age of the 'pill'. Mouse females produce a veritable birth-control scent. But it works much more precisely and delicately than the human pill.[11]

In sufficient concentration, this scent inhibits the development of the female sex glands. The more female mice there are living together, the more infertile they become. The scent of male mice can balance out this effect, as Dr. H. M. Bruce[12] has discovered. However, the scent must come from the mate. If a strange male is placed in the cage with a pregnant female, the male's scent puts a halt to the development of the embryos. Depending on how far they have developed, they are either reabsorbed by the mother's body or aborted.

Hence, marital infidelity is fatal to unborn mouse babies. But infidelity is also a degenerate symptom that always appears in periods of overpopulation, as it does among rats and squawker herons.

Scents as birth-control drugs seem to be very widespread in the animal kingdom. Their effect may be impressively observed among meal beetles.[13] These insects, which inhabit mills and granaries, reproduce rapidly. But as soon as their number exceeds two beetles to a gram of flour, the females devour their own eggs the moment they have laid them. What triggers this remarkable behavior is a chemical substance in the beetles' excrement. With increasing concentration the scent first diminishes the fertility of the females, then prolongs the duration of larval development, and finally causes egg cannibalism.

Every layman can experiment with the birth-control scent of tadpoles.[14] When a single larger specimen of these larval frogs is introduced into an aquarium containing a group of small tadpoles, the small ones incomprehensibly stop eating in spite of more than ample supplies of food, and soon die. In thirty gallons of water a large tadpole can compel six smaller ones to starve to death.

To further refine the experiment, one merely pours the water in which several large tadpoles have been swimming into the basin containing the small tadpoles. This alone will bring about the fatal lack of appetite. We may conclude that some chemical fluid is operative here. By this device nature guarantees a kind of priority to the first-born. Scientists have investigated the whole matter and made the most careful measurements. The quantity of excreted fluid, its dilution in water, and the effect of fatal lack of appetite are so balanced that no more frogs grow up in a pond than can later find enough food to keep alive. Here is truly an uncanny phenomenon. Imagine if something similar could be discovered for human beings! In the frog pond, at any rate, a population explosion is simply impossible.

This self-regulation of the density of an animal population provides the foundation of stability even as it fends off the dangers of overpopulation. Without such controls, the 'balance' of nature would not be worth much. For if that balance were perpetually unstable, storks, trout, and dragonfly larvae would be utterly dependent for survival upon the fate of the frogs.

Another experiment throws much light on this question. Freshwater fish regulate their populations in every lake in a manner similar to that of frogs. In 1965 a Canadian scientist who seriously questioned this principle attempted to provide a clear-cut demonstration to disprove it. Dr. W. E. Johnson[14] of the Fisheries Research Institute took a census of the fish population in a small mountain lake. Then he placed a non-native variety of trout in the lake. Trout are voracious fish which consume large numbers of small young fish. But after three years the other varieties of fish, in spite of heavy losses, were just as numerous as they had been in more peaceful times. The trout had eaten only the young fish which otherwise would have been eliminated by the birth-control scents.

Some sixty years ago African big-game hunters were astonished when they shot large numbers of elephants in one area, or many crocodiles along a strip of river, only to find that some years afterward the population of these animals had not perceptibly diminished. Today we know that basically the same phenomenon was taking place among these African animals as among the fish preyed on by trout.

The mechanism of birth control is different among elephants, however. Elephants are intelligent animals. When they realize that they are being exterminated in a given region, they do not let themselves be shot down to the last member of the herd. They leave the region and keep moving until they reach a place where they are unmolested. If they are lucky, they arrive at a national park. That is why these protected areas are subject to a veritable elephant invasion in times of severe persecution.

Serengeti Park in East Africa, for example, only three decades ago was reported to have no elephants in it. In 1958 Professor Bernhard Grzimek[15] in his great animal census counted some sixty of the big beasts. In 1964 the number of elephants in the park was estimated at eight hundred, and by 1967 the *East African Wildlife Journal*[16] recorded two thousand specimens. Murchison Falls Park in Uganda was actually overrun by ten thousand elephants in the same year.

It is scarcely surprising that in the first invasion of these elephant hordes, the stands of trees in some areas would be knocked down, uprooted, and destroyed. What is more surprising is that the animals quickly take measures to limit their numbers.

It throws light on the nature of elephants that in spite of overpopulation in certain places no symptoms of degeneracy in their social life have so far been discovered. Two Washington University scientists, Dr. Irven Buss and Dr. Norman Smith,[17] comment:

> Their young are exceptionally well cared for. They are kept in small 'nurseries' and guarded by communal 'governesses' whilst the affairs of the whole group are supervised by old bulls accompanied by vigorous young 'squires'. By contrast, overt sexuality, that is courtship and copulation, seems to be a relatively uncomplicated affair calculated to maintain the stability of the herd. Cows have multiple mates; there are no prolonged male–female relationships and frequently there is no fighting between rival bulls, although frustrated animals have been seen to roll on the ground, kicking their legs in the air like a baby.

The Murchison elephants have adapted to the new conditions by the simple procedure of having the cows wait longer between the birth of a calf and copulating again. Under normal conditions a cow elephant interposes a period of two years and three days between these events. Now she prolongs it to six years and ten months, more than three times as long.

We do not yet know what it is that prompts the elephants to such 'wise' continence. Still, may we be permitted to ask whether mankind in the future is going to behave like elephants or like rats?

Unfortunately, the self-regulation of population density in animal societies also has its dark side. That is apparent today everywhere in those parts of Africa and the rest of the globe where animals are defenselessly exposed to the depredations of man. If because of hunting, expansion of farmland, and plain extermination, a species falls below a critical limit in density of population, disaster descends suddenly upon the remaining stock. Usually the small remnant that is left dies out 'of its own accord'.

Scientists are at present facing this problem in the Galapagos Islands in the Pacific. Once upon a time hundreds of thousands of the giant tortoises that weigh up to half a ton lived on these islands. They were ruthlessly hunted by nineteenth-century whalers. Today there are only a few remaining specimens, which survived largely by chance, and these are slowly dying out. Since 1956 they have been under the protection of the recently founded Charles Darwin Station. The director, Dr. Roger Perry,[18] has had to admit that all solicitude for the last of the species is in vain. 'The giant tortoises no longer seem to have the will to reproduce.'

From these facts Dr. Schultze-Westrum[19] has drawn conclusions which he believes are valid for all vertebrate populations. 'As soon as a species reaches the threshold which leads from medium population density to underpopulation, and which is marked by the absence of the impulses or stimuli emanated by fellow members of the species, there is danger that the number of individuals will decline abruptly (for example by diminution of reproductive activity) and that the species in question will die out.'

But as long as the regulatory mechanism has not been wholly shattered, the animals of many species conduct virtual censuses and increase their offspring if they are too few or reduce the number of births if they are already too many. The concentration of birth-inhibiting scents among mice, tadpoles, and meal beetles can be regarded, according to Professor Wynne-Edwards,[2] as the result of such a census.

Animals have a variety of methods of showing their numbers—or making them audible. Examples are the nightly concert of frogs, the morning caroling of birds, the chirping of cicadas, the orgies of shouting practiced by South American red howling monkeys, or the interminable 'bub-bub-bub' of the redfish or croakers which spawn in schools of millions along the Atlantic Coast of North America. In Wynne-Edwards' opinion all these manifestations have psychological effects upon the animals involved and after a critical noise-level is reached the capacity to breed is diminished. Changes take place in the ovaries of the females leading to a growing infertility.

With other animals purely optical demonstrations serve the same purpose: the display flights of ducks at twilight, the 'practice flights' of squadrons of migratory birds, the dances of gnats in dense clouds, the ghostly city of light created by tropical fireflies, the mating ceremonies of the ruffs, hummingbirds, birds of paradise, and bowerbirds. The more forceful the impression of a mass, or of the superiority of others, upon the individual, the more that individual is psychically castrated.

Only a few years ago all types of birth control were regarded by many as diabolical and unnatural, a violation of divine law. But now science has shown that untrammeled nature presents more than rare, exceptional cases of birth control. Rather, the phenomenon is universal; hardly any species of animal can escape it.

To what extent, then, is man subject to these laws of nature?

Feeding Towers for Men?

If the human avalanche of reproduction continues to roll on in the future at the same acceleration, by the year 2040 there will be no less than twenty-two billion specimens of Homo sapiens scurrying about the crust of the earth—five times more than today! Man, biologically still in the Stone Age, still loving and needing an intimacy with nature, will be squeezed by urban planners into skyscraper barracks, compared to which our present cities will seem rustic havens. 'Our civilization is inexorably marching with flying banners from the laying battery for hens and the fattening box for calves to the standard feeding tower for man,' Dr. Paul Leyhausen[20] prophesies. Shall we survive such transformation?

If man possesses a biological regulator to prevent overpopulation, similar to that of social or gregarious animals, the mechanism is undoubtedly geared to conditions in primordial times. In those days infant and maternal mortality was enormously high. Everyone was almost daily threatened by death in the course of a hunt or a tribal feud. Men were helpless to combat diseases. The average age must have been far under thirty. Consequently, a biological regulator would not have been necessary and thus is hardly likely to be present today.

'According to the present state of knowledge,' writes Dr. Schultze-Westrum,[19] 'mankind cannot hope for the intervention of a natural regulatory system to reduce overpopulation.' The most beneficent of the great accomplishments of our century, the progress of medicine, is therefore Janus-faced, for it has ushered in the greatest danger to man in the whole history of the species: too many human beings! There are clear indications that man in the mass is unavoidably exposed to degeneracy symptoms similar to those observed in overpopulated animal societies. But—and this is the terrifying aspect of the matter— these degeneracy symptoms will scarcely have any regulatory effect upon humanity. They can, however, lead directly to disaster for man as a species. From overcrowding to total nothingness is only a step.

Almost five years as a prisoner of war taught Dr. Leyhausen that overpopulated human communities exhibit the symptoms of over-populated wolf, cat, goat, mouse, rat, and rabbit societies down to the last detail. Aside from peculiarities typical of each species, Dr. Leyhausen concluded, the essential motive forces are in principle identical.

The first controlled experiments with closely packed masses of human beings seem to confirm these conclusions by analogy. In an Oxford children's clinic Dr. Corinne Hutt and Jane Vaisey[21] divided children into three groups according to their dispositions. Each group was separated, and at first only a few in each were placed in relatively small playrooms. From day to day a steadily increasing number of children were added to the three rooms. The boys and girls were not supervised, but were under constant observation through one-way panes of glass.

As the crowding in the room steadily increased, the children with well-balanced (normal) temperaments withdrew more and more into themselves. They increasingly avoided contact with their fellows. The more jammed the room became, the more they isolated themselves. But when the crowding became so great that the children could no longer avoid physical contact with each other, aggression flared up everywhere and the experiment had to be ended. The character of these children in their reaction to crowding might be called similar to that of elephants.

In the group of children with quarrelsome dispositions (caused by brain damage), almost continuous bickering began as soon as the room reached a medium degree of crowding. That is the ratlike reaction. The third group (composed of autistic children) seemed to enjoy remaining passive even in the midst of the densest crowd. We call this the locust type.

Among human beings there are far more shadings than in the behavior of animals. This makes the problem more complicated among us. The human mass psyche—an utterly irrational phenomenon—is not a unity. Each individual reacts in the mass somewhat differently, according to his personality.

Here is a highly illuminating example. In baboon society in the wild[22] there are no despotisms. There does exist a carefully negotiated ranking order which is not easy to recognize because it is complicated by friendships, alliances, male leagues, and female cliques. But the fine structure is well knit. Within a definite framework it allows every ape an area of personal freedom. Every male baboon has his own proportionate right to food, to a sleeping place, and to access to the females. The 'boss' is never the absolute ruler of a harem for himself alone.

But in the close quarters of confinement in a zoo he is exactly that, to an intolerable extent. Here the fine structure of the social forms breaks down. The baboon society then consists of a tyrant in the most obnoxious sense of the word at the top, and one or two 'scapegoats' at the very bottom, while the group between these two extremes forms an amorphous, socially unstructured mass. Paul Leyhausen observed the same development when he kept many cats in close quarters. He applies his observations to man in the following terms: 'For man, too, overpopulation represents a danger to genuine democracy. The result is almost inevitably tyranny, whether this is practised by a personal tyrant or an abstract idea such as the commonweal—which for the majority of the citizens results in more burden than profit.

'There appears to be an unalterable law operating here. As long as the density is bearable, sacrifices for a common cause somehow contain their reward within themselves and thus help to give meaning and also happiness to the life of the individual. But if the density of population increases beyond the tolerable limit, the demands of the community mount steeply, and what is taken from the individual is lost to all.'

Present-day prosperity in Europe and North America somewhat masks the fact that whole nations are already living in a state of overpopulation. Unfortunately, the first warning symptoms are being misinterpreted. Existential anxiety, fear of being crushed by the massive competition, has already made the heart attack the principal cause of death. The pathological fear of examinations experienced by students condemned to vastly overcrowded classes has a similar psychic genesis, as have 'hypochondriacal depressions' which trouble the uprooted Italian workers[23] in Germany.

No statistics have been gathered on those far too small apartments in which parents driven to neurosis create an atmosphere of endless whippings. For the hapless children such homes are simply breeding grounds for future criminals or miserable neurotics. Everyone knows how flimsy walls in an apartment house produce antagonisms between neighbors—not because 'the others' are outrageous people, but because without isolating walls we have a situation like that of the kindergarten experiment described earlier.

One of the ingrained peculiarities of man is that he is fitted for social life only in a small group. In order to combat anxiety he needs the shelter of a community whose dimensions he can measure and in which he has his fixed place. Too much company—in other words, the great anonymous mass—robs him of this feeling of shelter and safety. The inexplicable anxieties return. He reacts with all sorts of complexes, blocks, repressions, aggressions, and fears, which soon grow into outright neuroses. Nowadays we label these anxieties fear of the atom bomb, of the Communists, of the Chinese, of overpopulation, of the rule of robots, of losing one's job to automation, of incurable disease. Why should 'the Establishment' wonder that youth resorts to revolutionary excesses in seeking a way out from these all-encompassing uncertainties? The only alternative is resignation and apathy.

'We must learn from the behavior of animal collectives,' says Professor George M. Carstairs,[24] the Edinburgh psychiatrist, 'that there are natural limits to a population which cannot be crossed with impunity.' Science, he continued, is not yet in a position to predict what may happen to mankind in different phases of overpopulation. But in any case the danger of a complete demoralization of the masses is already present. If old social structures collapse, powerful irrational forces will gain ascendancy in the human collective. And nowadays a burst of irrationality can easily lead to total catastrophe.

To choose an example straight from the present: the general feeling of insecurity creates an intense subliminal striving for security. So much is quite understandable. But since the responsible leaders of the world are not aware of the deeper biologic causes of this striving, they give it a wrong turn. They see security as a military matter and take measures accordingly. The result is that in both the East and the West the stockpile of insecurity grows, in the form of arsenals of nuclear weapons sufficient to transform the whole earth into a cratered landscape like that of the moon.

Who today, aside from a few scientists, recognizes the real problems humanity must solve if it is not to exterminate itself? And who is ready to tackle these problems? For dealing with overpopulation and the consequences of automation will take considerably more money and imagination than the insane atomic arms race.

A cynic might remark that no one ought to hope for so much money and, above all, for so much imagination. The most rational 'final solution' of the population problem, he might point out, is the hydrogen bomb. After all, that is how things happened in the animal world, wasn't it? When lemmings overpopulate their territory, they drown their numbers in the sea. . . .

Do Animals Commit Suicide?

Only a year before, hardly a lemming could be found in this corner of Alaska. But now, on a warm spring day in 1967, the earth seems to be teeming with them. By the millions and tens of millions these burrowing rodents, which look rather like hamsters, emerge from the ground. Anyone walking in the area will trample on squeaking lemmings at every step, no matter how he tries to avoid them. As if seized by mass madness, the animals form into a host that stretches for miles. And now the horde rushes in an undeviating straight line across the Alaskan tundra, crossing mountains, swimming rivers and lakes.

After a march of over 125 miles the vanguard reaches the precipitous coast near Point Barrow. But still it does not stop. Blindly, the endless ranks plunge down from the cliffs into the cold surf of the Arctic Ocean. At first they swim for a while. Then the cold numbs their little bodies. The ocean becomes the grave for hundreds of millions of the animals.

The lemmings' suicidal rush into the sea has hitherto been regarded as almost the only example of a mysterious death instinct among animals. In northern Norway the legend arose that the lemmings were obeying an instinct to migrate to sunken Atlantis, or even to Greenland, which was supposedly connected with Scandinavia in former geological eras. In northern Sweden and Finland the story goes that the animals are making for a mountain which in earlier times occupied the area of the present Baltic Sea. But the favorite explanation has been that the surplus population was deliberately sacrificing itself; that after years of explosive reproduction the lemmings reduced their numbers by mass suicide in the sea.

Africa used to present a somewhat comparable phenomenon. Before the white man with his rifles thinned out the great herds, lemminglike mass increases used to take place from time to time among springboks.[25] Migrations of up to fifty thousand of these gazelles would head for the Namib Desert in southwest Africa, where they died miserably. Sailors have also sighted swarms of locusts over the middle of the Atlantic,[26] two thousand miles from their birthplace in Africa. But the insects' stamina would give out and they would drop by the billions into the sea.

But are such phenomena really suicide as we human beings understand the term? Not at all. For the locusts were bent on finding new land—'like Columbus'. But in these mass flights they are at the mercy of the wind and, when driven westward, have nowhere to alight. The springboks had exhausted their usual grazing grounds and were desperately searching for new pastures. The case of the lemmings is rather similar.

What causes some animals to multiply so excessively in three to four year cycles? In northern Scandinavia the most recent lemming years were 1960, 1963, and 1967. At Point Barrow on the northern coast of

Alaska the years 1946, 1949, 1953, and 1956 were times of incredible lemming increase. Dr. F. A. Pitelka,[27] who studied the phenomenon, reported that, after the gathering of the lemmings, a coastal strip two hundred miles long and fifteen to twenty miles wide looked as if it had been 'mowed down'. More animal bodies than grass could be seen.

A procession of millions of lemmings cautiously crossing a lake.

Certainly we are not exaggerating when we speak of a population bomb in connection with these little rodents. The greater the population density becomes, the larger the litter each female bears and the more rapidly the succession of litters. The gestation period of lemmings is only twenty days. The young are sexually mature at the age of twelve days, and the females can copulate again within a few hours of giving birth.

Here, then, the birth-control mechanism as we have come to know it in the case of mice, rabbits, elephants, and many other animals, has been perverted into its exact opposite. The effect of masses does not brake reproduction, but accelerates it beyond all bounds—as is also the case with locusts.

Even taking generous estimates of initial population figures, the scientists were unable to arrive at any reasonable reckonings. Each time the population explosion occurs, only a few 'abnormal lemmings that remain normal' survive by not joining in the general exodus. It would seem impossible for these few survivors to increase to the point of once more flooding the country with lemmings in anything less than ten years. Yet the fact remains that within three or four years at most new swarms of these overpopulation fanatics have appeared.

How is it that the lemmings actually reproduce faster than calculations predict? In old books of natural history accounts are given of lemmings suddenly raining down from heaven by the millions. And in our own day some investigators quite seriously posit that the lichens on which the lemmings subsist are from time to time enriched by some mysterious fertility vitamin.

The reality, however, is quite different. The Finnish zoologist Dr. O. Kalela[28] has discovered that lemmings possess a capacity that is probably unique in the animal world. They reproduce not only in the summer, like all 'reasonable' animals, but also in winter. Hitherto it had been assumed that lemmings quietly hibernate in the burrows which form their winter quarters. But in fact their avalanche of offspring does not stop even in winter. That is why lemming years are so hard to predict. One fall there will be still no sign of overpopulation. But the following spring endless lemmings suddenly pour out of the burrows, until it almost seems that pure mud has been transformed into living creatures.

The reproductive activity of the lemmings is interrupted for only two weeks in the year: at the time of the two seasonal migrations in spring and fall. These movements between summer and winter quarters take place independently of the population density. In other words, they also occur in 'normal' years. But until 1960 these movements went unnoticed because the animals migrate secretly and individually by night. In 1960 Dr. Kalela discovered that this change of place was the key to the lemming mystery.

In summer the Finnish lemmings occupy moors high in the mountains, near the timber line. But in winter these areas freeze hard, and the rodents therefore must move to their winter quarters before biting cold descends. They find refuge either in places above the timber line where dwarf shrubs offer protection, or in pine forests below the timber line.

The migration of individual animals, or at the most small bands, undergoes a fundamental change after the usual two- or three-year interval, when mass reproduction has taken place in the underground winter quarters and hordes of the animals pop out of the ground in the spring. Then they quickly overwhelm and eat bare their usual summer estates. Very slowly at first, the dense population expands downhill into less suitable areas. The animals assume extremely aggressive attitudes toward one another. Fights break out incessantly. But few serious wounds are incurred because lemmings possess a thick leathery armor under their soft pelt, protecting the vulnerable areas of their heads and bodies.[29]

The expansion into leaner territory soon ends at some small brook. Masses of lemmings are checked here. Bickering and fighting increase madly. Eugen Skasa-Weiss comments: 'To the individual lemming the whole world must seem intolerably overlemminged.' All at once mass psychosis descends. Millions of 'refugees' race about in wild confusion. Suddenly one small band crosses the brook. The others follow. Then a mile-long procession forms, advancing at a steady run. Charles Elton[27] of Oxford has observed columns up to 125 miles long.

The direction taken by these columns is purely arbitrary, and depends solely upon the accidental course taken by the lead group after the crossing of the brook. But once the direction has been chosen, the column keeps to it. We do not know what means the leaders use for orientation. Nature seems to have given lemmings an internal compass that keeps them on a forward course. But they certainly do not aim for the seacoast 'in the grip of some mysterious suicidal impulse'.

The small rodents are good swimmers. They can cross rivers and lakes even miles in width. But when they come to a body of water, they do not, as they are reported to, plunge into it blindly. Rather, they change their otherwise straight course and run up and down the shore looking for a flat beach. Then they move into the water with extreme caution. But if they find themselves on a precipitous coast with no favorable spots for entering the water, hunger and migratory madness will leave them no choice. Then, but only then, do they take the leap as if committing an act of extreme desperation.

The frenzied march of the lemmings resembles the retreat of Napoleon's army from Russia. Exhausted stragglers remain lying by the wayside. The brooks and rivers they have crossed wash away the bodies of the drowned by the tens of thousands. Only on rare occasions do the restless migrants have the good fortune to come upon new territory suitable for their needs, where they can settle. What is more likely to happen is that in such regions they once more encounter a mass of lemmings, carry the new hordes with them, and so swell the column marching to its death. Even if they do not reach the coast, and this is often what happens, the millions die gradually of stress, exhaustion and hunger.

In distinctly rarer cases chance leads the obsessed travelers to the seacoast. Here the nearsighted animals are evidently incapable of distinguishing the open sea from a wide river or a lake. Their powerful migratory urge drives them onward. The lemmings swim in dense squadrons farther and farther out into the ocean, until after several hours of plucky kicking and paddling they drown.

Their sad fate is thus the consequence of an instinctual, hopeless search for living space, not a 'mystical, suicidal mass delusion'. There is no death instinct involved, nor any 'inner knowledge of the right way to the Beyond', as some have sermonized. But surely it is far more extraordinary that the lemmings should provide us with a perfect example of the catastrophic madness to which a population explosion can drive blind and maddened masses!

Perhaps, however, we can find other examples which do suggest the presence of a death instinct in animals, examples of suicide, or at any rate examples showing that animals in hopeless situations prefer death to a terrible future.

Hunters in the high mountains have often told stories of pursuing an ibex to the edge of an abyss. The animals, they say, invariably choose the fatal leap into the depths. But that surely has nothing to do with a morality of preferring death to slavery. Fear of the hunter was simply greater than fear of the uncertain fall into the gorge.

Many human suicides likewise take the path of no return from an extremity of anxiety feelings: fear of some general doom, a dread of economic ruin, or of losing a sweetheart, or of a father's severity, or of difficulties at work.

It is highly questionable whether an ibex immediately before the leap exercises any judgment. Since the ibex is impelled only by fears and has no conception of the meaning of death, his behavior cannot be called suicide. On the other hand, most persons who take their own lives are the prey of a deep depression with and without cause. There is indeed a sickness of the soul, a longing for death. Something comparable can be found among animals.

Dr. Erich Baeumer,[30] for example, recounts the tragedy of Audax, his once-proud cock. In the glory of his splendid plumage, Audax dominated the hens and cockerels of his barnyard for several years. But by and by one of his strong 'sons' grew up. As is the way of the world, this bird attempted to depose his father. The result was a fierce cock-fight in which youth won out.

Audax, however, refused to recognize his defeat. Next day he tried to undo his disgrace by fighting again. In vain. On the third day also he failed to reconquer his domain. Instead his successor chased him about the whole yard and thrashed him whenever he came into view. At last Audax surrendered. To avoid more blows, he always had to crouch when his lordly son passed by. He had to lift his wings slightly and by twitching his shoulders raise the white flag of surrender, so to speak. Nor was he allowed to crow any more; the most he dared to do was to cackle softly like a hen.

The splendor of his plumage faded. Audax became thin, scraggly, and filthy. After two weeks he died, of no visible cause.

Grief can weigh so heavily upon an animal that it no longer can muster the strength to continue living. We are familiar with this phenomenon in dogs that are extremely attached to their masters. They do not long survive the deaths of their masters or mistresses even if they continue to receive the best of care and are in the full flower of their strength.

Heartsick animals, however, always decline slowly. A melancholy dog never deliberately throws itself in front of a moving car. A cock weary of life never deliberately puts itself into the path of a savage chained dog. At the decisive moment fear of danger is greater than weariness with life. That is the difference between animals and men.

The point we have already made remains valid: neither cock nor dog has any conception of death, let alone of a hereafter—such as religion and philosophy or our own imaginations confer upon us. The phenomenon of suicide is inseparably linked with such conceptions. Man is the only organism who can put an end to himself with full knowledge of what he is doing.

And here the matter would rest unless there should be another creature on earth who has an inkling of the meaning of death.

If there is such a creature, it is probably the chimpanzee. Dr. Adriaan Kortlandt's experiments in the Congo jungle, which we have described in our first chapter, suggest the possibility of such insight. We will recall the chimpanzee's fear of death expressed in his drawing back from dead animals, from the arm or leg of a member of his own species, even from sleeping animals and lifeless images. Such fear is of an utterly different nature from the fears manifested by other animals. Even rhesus monkeys are not at all affected by the sight of a beheaded fellow.

But whether chimpanzees with this degree of consciousness of death are ever impelled to suicide, no one can say at present. No such case has, at any rate, been reported so far. On the other hand, the apes have not been observed over sufficiently long periods of time to justify any final opinion.

It may also be that chimpanzees have a much stronger fear of death than men and for this reason alone are incapable of suicide. To the superficial view, the anthropoid ape's fright reaction at the sight of a corpse is certainly far more violent than ours. In this sense these animals evidently see the situation far more realistically than a man entangled in delusions and emotional confusions who wants to take his own life.

In these pages we have talked a great deal about fear, about the various forms of fear, the mysterious imponderables of the emotions, and about mass hysteria and panic which lead to disaster. But what is fear in itself?

Fear Can Be Trained

There are people who are only 'happy' when they are unhappy, when their nerves are keyed up by danger, when they are burdened by oppressive debts or involved in the complications of a broken marriage. If these troubles are lifted from them, they fall ill. Or else—as Dr. Eric Berne[31] has noted in his *Games People Play*—they deliberately seek out the nearest available situation of danger, or new financial burdens, or another unhappy marriage. Dr. Paul Leyhausen[32] has attempted to clarify these contradictions in the human psyche, with which psychotherapists must constantly deal.

His initial investigation of what fear really is led him to some surprising perspectives. Men are all too easily inclined to associate fear with some presumed cause: fear of death, of atomic war, of a traffic accident, of a loss. But paradoxically, the sense of fear can exist irrespective of any actual or imaginary cause.

Most animals can feel intense fear without the slightest inkling of what they are fearing. A mouse, for example, fears an open area without cover by day just as much as a man does a dense forest at night. Yet the mouse is not trembling for fear of any specific predator. It exceeds the capacity of a mouse's brain to reason that under these circumstances a cat has better chances of catching it. Even young laboratory mice that have never seen a cat, fear all areas that are too bright and open.

If a mouse had to learn by experience that a cat might be present on a bright, flat surface poor in cover, that it could easily be caught under such circumstances, and that being caught meant death—the mouse would be dead after that first experience and experience would do it no good at all. A banal enough observation. Yet it is the key to understanding the extremely involved phenomenon of fear.

Nature has given her creatures an 'unconscious knowledge' of the typical characteristics of dangerous things and situations. A mouse automatically suffers a kind of agoraphobia on a bright open surface. It becomes afraid as soon as there is a rustling in leaves which does not sound like the rustling of a mouse. It becomes afraid at the mere sight of a dark something slowly circling above it (predatory bird pattern) or when it hears the alarm call of another mouse. This stereotyped triggering of the fear mechanism proves far more useful to the animal than any vague knowledge that it must die combined with experiences of all the perils that may lead to death.

According to Paul Leyhausen, the perilous position of animals, who would all meet untimely deaths were there no instinctive fear to protect them in dangerous situations, has been a goad to natural selection to develop many forms of fear reactions, avoidance behavior, and flight patterns. But mortality itself (this is the paradox) does not appear to be the motivation for feelings of fear in animals, except for the chimpanzee.

The original reaction, then, is creature fear, which knows nothing of death or of the nature of danger. Only man, and presumably the common ancestor of man and chimpanzee, has thoughts about fear and intertwines rational thoughts with natural feelings.

Nevertheless, man himself is not free of creature fear. Fear is innate in children. They are afraid before instruction and their own experience teach them the connection between fear and death so that they can actually fear death as a secondary phenomenon. That is why fear contains that irrational, a priori element concerning which many a philosopher and psychoanalyst has been led to both brilliant and bizarre speculations.

Fear, then, is primarily an instinct. As such, it is subject to the laws which Professor Konrad Lorenz[33] has presented in connection with aggression. An instinct is an operative pattern of nerve combinations and hormonelike exciting agents. Such innate dispositions wait, as it were, to be triggered by a specific outside stimulus. They then react by producing an appropriate quality and quantity of emotion and a mode of behavior that is (usually) biologically meaningful.

In other words, the body literally produces fear independently of environmental conditions. The hormonelike exciting agents are constantly forming and accumulating, so that they will be available in time of need. The longer an organism has had no fear, the more the fear-exciting substances accumulate, since they have not yet been used up in fear reactions, and the more trivial the causes that can trigger fear.

Ultimately the creature is driven to search for a situation in which it will certainly be fearful and can therefore discharge the stored impulse.

At first glance it may seem incredible that an animal or a human being can actually have a kind of longing for fear. But who has never played with fire, engaged in a dangerous sport, or provoked a quarrel? Who has never devoured a detective story, shuddered at a horror film, dared some reckless maneuver in traffic? Why do children play cops and robbers and enjoy riding on roller coasters? And why does the military life constantly gain attractiveness as a lost war recedes into the past?

Fear, even the most fearsome fear, is interwoven with the attractions of pleasure. In extreme cases it can become a mania, a neurosis, or what is often mistakenly called the instinct for self-destruction—which does not exist at all. The only question is: where does the natural end and where does morbidity begin?

'In the course of evolution nature has instilled in every species of animal a production of fear corresponding to the average degree of danger encountered,' says Dr. Leyhausen. There are so-called animal paradises, islands in the Pacific and Indian oceans where since primordial times there have been no large land predators and therefore no danger to some of the animals. Consequently they lead a life without fear. The buzzards, the iguanid marine lizards and the sea lions of the Galapagos Islands[34] fearlessly let men approach them, exactly like the Antarctic penguins. The lion, too, is by no means an especially courageous animal, as is generally thought. He merely has few enemies and therefore little fear. Lack of fear is not the same as courage.

On the other hand, the penguin, which fears nothing on land or coming out of the air, is terrified of being the first of a group to jump into the water. His fear is of a highly specialized sort; its purpose is to protect him from enemies in the water, seals, sea leopards, and killer whales.

When gallinaceous birds hear their species signal meaning: 'Beware! Ground enemy!' (fox, cat, wolf), they flee without having seen the enemy. Their fear alone sends them fluttering up to a tree. But when the alarm for raptorial birds is sounded—it is an entirely different call—

25 (RIGHT) *Bloodless bout between two mice. Members of the same clan never bite one another.*

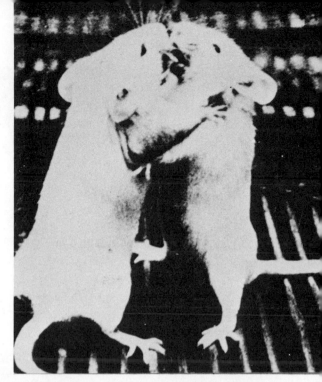

26 (BELOW) *Peaceful idyll in the hamadryad baboon family. The mother (middle) tends her baby with the help of an 'aunt'.*

27 Baboon mother training her baby. To keep the precocious little fellow from rushing into dangerous adventures, she holds him by the tail like a dog on a leash. The mother also shows the baby what he is allowed to eat, what unpalatable things to avoid, and how to behave toward others in refined ape society.

28 With his mother as protector, Wastl, the four-week-old saimiri baby, postures toward an adult member of the troop (not visible in the picture).

29 (LEFT) *Ring-tailed lemurs in the jungles of Madagascar lead a community life that demands high social intelligence.*

30 (BELOW) *Lemurs like this Ceylonese loris cannot grimace because they lack the facial musculature.*

31 The beaver learns hydraulic engineering from his parents. His dams are technically perfect. The lodge is inaccessible to enemies; the winter food supply is stored under the ice in just the right amount. The beaver toils day and night to maintain his living standards.

32 (ABOVE) *Two animals that use tools. The Galapagos woodpecker-finch pokes insects out of rotten wood by holding a stick in his beak.*

33 (BELOW) *The beaver cuts a felled tree into transportable sections to support his dam.*

34 *What a friendly laugh this chimpanzee has, we are tempted to say. But we can go badly astray in interpreting animal facial expressions anthropomorphically. Opening the mouth wide, showing upper and lower teeth, holding both hands to the head and tugging at the ears—all this may be a sign of extreme fury in the chimpanzee, or a sign of grief. One should be very careful of a chimpanzee who acts this way.*

35 *Two stags tangle in the Harz Mountains. Fighting with antlers never takes place in the spring, when the antlers are coated with velvet and highly sensitive to pain. The stags therefore rise on their hind legs and use their forelegs like boxers. This is one of the few cases in which animals fight in erect posture.*

the fowl scurry quietly into cover on the ground. Thus it is probably wrong to speak of a single instinct of fear. Presumably an organism harbors several types of fear.

The marine turtle also has at least two kinds of fear. If you tap on its shell while the turtle is on land, fear accelerates its heartbeat. In the water the reaction is reversed. There fear slows its pulse to one or two heartbeats per minute. We do not know whether the turtle also experiences two different sensations of fear. It is quite conceivable that both fear instincts run to a kind of 'common terminal' in the nervous system. Is it possible that man also has several kinds of fear which he can ill distinguish?

Even shattering anxiety states can be quite normal, hence not morbid. Typical 'prey' such as antelopes, mice, and songbirds, constantly live with fear. In all their actions fear is a permanent concomitant. Such animals can more easily afford to go without food for several days, or to miss a chance for copulation, than to let their continual tense alertness slacken for even five minutes. A deer sleeps only two hours a day, a giraffe a mere seven minutes. Those who insist that 'hunger and love' dominate the world are certainly not too well acquainted with fear.

In man the average production of fear agents is geared approximately to the dangers that existed in the Stone Age. Since that time man in civilized countries has mitigated or eliminated many sources of danger: large predators, the need for daily hunting, endless tribal wars, blood feuds, a great many diseases and bodily afflictions, lack of shelter from the weather. Electric light has also robbed the night of many of its terrors.

But the production of fear agents within man's body has remained essentially the same, although there is a considerable range of variation for individuals. This fear must find an outlet. Thus man in the shelter of gregarious life has found substitute objects which he can use for the discharge of his superfluous fear. His imagination has invented superstitious fear of demons. 'A ghost is the projection of a now nonexistent predator,' says Konrad Lorenz.[35]

'Seeing ghosts' also occurs in a figurative sense. If someone is seeking triggers for his fear, he falsifies innocent acts in order to give them the desired interpretation. He attributes some evil intent to his fellow man in order to be able to fear him. If this phenomenon takes extreme forms, it is called paranoia. Another device for triggering fear leads to hypochondria. Woe to the doctor who fails to find some illness in the hypochondriac. For the reasons given here such pseudo-patients crave confirmation of their phantom illnesses in order to be able to feel really afraid.

But there is a deeply troublesome aspect to the whole phenomenon: the stored fear is not consumed solely in feelings of terror. For that is only the emotion which normally inspires avoidance or flight behavior. Only action of some sort can produce the needed discharge. According to Paul Leyhausen, when the normal motor response to fear is blocked, when flight becomes impossible for one reason or another, fear then activates another instinct, its great counterpoise: aggression. If this recourse is barred to man or animal the undischarged fear leads to illness.

Here is a dramatic example. In a laboratory experiment Dr. Jules Masserman[36] continually confronted a rhesus monkey with insoluble decisions. The animal was given a choice of two levers to press. He had to press the right one to receive a small reward of food. If he chose wrongly, he received a mild electric shock. But the experimenter alternated rewards and punishments in such a way that the monkey could not discover which lever was the favorable one. The animal was made to 'work' eight hours a day in this fashion. Within a short time it developed an 'experiment neurosis'. Its states of anxiety mounted steadily, since it could neither flee nor attack anything. After a few days it developed high blood pressure and all the symptoms of human 'executivitis'. It died of a heart attack.

A controlled animal which received the same number of electric shocks, but was not made to inflict these on itself by wrong decisions, remained healthy and cheerful. The mild electric shocks, therefore, had not affected the second monkey's health. But the first monkey's fear of making the wrong decision by pressing the wrong lever cost him his life.

The same thing happens to a human being who must 'press levers' and never knows which one is right—either because he lacks the capacity to master his task or because his superiors or partners react unpredictably. Everyone who repeatedly finds himself in such blocked situations, either actually or only in his imagination, will either suffer from 'executivitis' or gradually grow more and more aggressive—or both. Since his unnecessary contentiousness involves him more frequently in dangerous situations, the internal secretion of anxiety increases also. Thus the vicious circle is formed.

These findings do not call into doubt the instinctual character of anxiety. For an instinct is not rigid and unalterable; that is a long outmoded view of it. Every instinct has a spectrum of variation. Instincts can be weakened by habituation, for example. That is what happens to deer in preserves, where they show only mild fear-reactions to the presence of human beings.

An instinct, however, can also be trained; it can be incited to a point of neurotic excessiveness. Then it begins to operate morbidly and destructively.

This process takes place most devastatingly in a human being when it begins in early youth—especially under the influence of parents who whip their children from the time they are small for every trivial fault. It is a probability verging on a certainty that these children who have morality beaten into them will in later life become either extremely neurotic or incomprehensibly criminal. But of course only those who know nothing about the biological laws of fear will find such development incomprehensible. For what the parents have done instead of teaching right and wrong is to cultivate anxiety, sweet anxiety. The child begins being a nuisance at home solely in order to secure 'his' whippings—or more precisely, his fear of the whippings. Later, as an adult, he will become a criminal not because man is by nature evil but because he has become addicted to that whetting of his nerves.

This is so near to being a natural law that predictions can be made on its basis. In 1954 Sheldon and Eleanor Glueck[37] of Harvard University drew up psychological 'horoscopes' of 303 six-year-old New York slum boys. Judging solely from the manner in which these boys were raised by their mothers, the two scientists predicted that thirty-three of the children would develop into criminals. The prognoses were kept secret, of course. Ten years later twenty-eight of the boys, by then sixteen, had already come into conflict with the law for a variety of crimes and misdemeanors.

Persons subject to overproduction of anxiety are plagued by nightmares in their sleep, and by day seek some masochistic counterpoise. This is true whether their excess of anxiety is constitutional or implanted by training. Professor Rudolf Bilz[38] of Mainz speaks of a patient 'who shudders when he merely hears the word "thunderstorm" in the weather report. Yet this man says: "But in the war I felt completely free of fear." He didn't have to shudder then, of course, for he was in the midst of the thunderstorm.'

When a victim of neurotic anxieties is really exposed to frightful situations, in war or in a concentration camp, he no longer fears. But as soon as the war or the ordeal is over, the anxiety returns—and returns moreover, with the same force that gripped normal people in major battles or in Treblinka. This fact gives us some measure of the intensity of the anxiety neurotic's suffering in peaceful times.

On the other hand, psychiatrists are also familiar with many cases in which perfectly normal persons first learned the true meaning of terror, in the most literal sense of the word, in concentration camps, and as a result of that experience became pitiable anxiety neurotics. Anxiety, then, can be implanted even in adults. The idea that anxiety is not a purely psychic process but also a biochemical event—the production or consumption of anxiety-exciting substances—is only a postulate, growing out of Konrad Lorenz's theory of the instincts. But Professor Ferris Pitts and Dr. James McClure,[39] working at Washington University in St. Louis, actually discovered this substance in 1967.

Two minutes after they injected lactate, the salt of lactic acid, into anxiety neurotics, grave attacks of anxiety began and continued for twenty minutes. For the first time anxiety had been artificially and predictably produced by a chemical stimulant.

The effect could be considerably weakened or even quelled by simultaneous administration of calcium ions. The two psychiatrists stress, however, that it is still too early to hold forth any hopes that anxiety neurosis may be treated by chemotherapeutic methods.

To a certain degree men are able to liberate themselves from severe anxiety by their own efforts. They need only realize clearly that they are suffering from anxiety, and what anxiety is, and that in the given situation they perhaps need not have it, or have it in lesser measure. Then they can rise above their anxiety and more or less control it. Hence the most frightful of our daytime fears is not as appalling as a nightmare which cannot be tamed by reason. Victory of the mind over the body does generally lie within our powers.

The phenomenon of fear is inseparably linked with its opposite: aggression. Quarrelsome people are usually very fond of fear (not fearful!). There has long been a great deal of discussion of man's supposed delight in evil. Of late scientists studying animal behavior have achieved new insights into this ancient problem. They have learned, for instance, that animals hitherto classified as extremely aggressive are not so at all. Let us glance at some examples.

5. Statesmanship Among Wild Animals

The Friendly Beast

A mighty, golden-maned lion trotted through the grass of the African plains not a hundred yards to one side of a herd of peacefully grazing impala. The graceful creatures were apparently totally uninteresting to him. The antelopes seemed to take this for granted. They did not think it necessary to flee, and calmly went on nibbling at the grass.

A quarter of an hour passed. Then, somewhere in the plains, a sound like a cough could be heard. The lion's ears twitched and, as if shot from a catapult, he rushed toward the antelopes. The herd instantly scattered in wild flight, much too fast for the 'king of beasts' to catch up with them. The lion promptly dropped back into his amiable trot. Obviously he knew what would be awaiting him in the nearest hollow between the grassy hills: his mates—with two slain antelopes between their paws.

These lions worked together, following cunning, carefully planned tactics. The lionesses had crept into position against the wind, unnoticed; then one of them signaled that they were ready. Thereupon the lion, who had only been there to distract the antelopes, drove the unsuspecting quarry to their destruction.

The plan and the execution were perfect. Probably Stone Age men could not have done much better. Each must be able to depend on the others. Lions and men both must be able to rely on their fellows, however they may appear raging beasts to outsiders.

Hitherto we have known little more than the savage side of the big cats, for we have relied on the stories of big-game hunters. But in recent years zoologists have been investigating a multitude of astonishing details of the tender family life of lions. The results of their work has considerable relevance to any discussion of the 'predatory components' of human character.

In 1961 the English game warden Norman Carr,[1] stationed in the Kafue National Park in northern Rhodesia, made extensive observations of leonine courtship. Each morning a lion with a magnificent mane would come calling on the lioness of his choice. If on occasion he slept too long, she rubbed her sinewy figure provocatively against his flanks. But as soon as he attempted intimacies, she drew back, hissed furiously, and gave him a few clouts. In other words, she systematically tormented him according to all the techniques of femininity.

To the human observer such premarital bickering looks quite murderous. In fact it is a rough form of tenderness that is customary among lions. It is as much part of the courtship as coyness is with us. These spats never develop into serious quarrels. And in contrast to human beings, neither of the two lions has any fear that they will. For all through the episode each is assuring the other by nuances of facial expression that the rebuffs are not to be taken seriously. They are amicable gestures corresponding to smiles among human beings.

Finally, after several days of ritual courtship, the lioness submitted to her suitor. Henceforth united in all things, the pair separated from the rest of the pride. A newly established pair of lions usually takes a honeymoon of several days. Their love play can be neatly described by the old phrase, 'I love you so much I could eat you.' Impetuously, the male lion takes his lady's entire head into his enormous maw. It looks as if he intends to devour her. But he only nips and scratches and licks her with the greatest tenderness.

The honeymoon is a most dangerous period for a boss lion. By lion custom, this is the great occasion when he may be dethroned. Sometimes a male rival decides that his chance has come. He creeps covertly behind the couple and usually forces matters to a fight. It begins with an attempt on the part of both males to howl each other down. The sounds they produce range from low growls and barbaric snarls to nerve-shattering roars that reverberate for miles across the plains.

If this contest in minstrelsy fails to establish dominance, the blows begin. The opponents belabor one another with slaps of such force that they would break a man's neck. But male lions possess a well-padded 'fencing mask': their shaggy manes. These are not just brave ornaments; above all they are shock absorbers. Moreover, in fighting with members of the same pride the two rivals always keep their knife-sharp claws withdrawn into the pads of their paws. Thus the fight, for all the enormous muscular power involved, is a thoroughly fair and bloodless tournament. So much for the human notion of lions as 'brute beasts'.

If the boss lion wins, everything remains the same. The defeated rival skulks back to the pride, and when the boss turns up again a few days later, neither shows signs of harboring a grudge. But if the challenger wins, the pride of up to twenty members may break up. That depends entirely on how much sympathy the challenger has succeeded in winning beforehand. Here too, then, we encounter a variety of primitive democracy which has its biological purpose. The larger the pride, the more difficult it is for the challenger to win over all the members. Sometimes the single pride splits into two parts.

If, however, the new boss of a smallish pride succeeds in winning all the other lions over to his cause, the dethroned male is sent 'into the desert'. Henceforth he must go his own way as a solitary. In contrast to an established pride, which in the African plains will defend a fairly definite territory of about one hundred square kilometers against other groups, the expelled lion henceforth has no home. He must become a wanderer.

That is a very harsh fate. For as soon as he enters the territory of a band of lions, he is chased away. And in the open plains of Africa, the territory of one pride borders directly on another. Thus the unfortunate loner is harried on and on until he reaches such parched regions that rats, mice and lizards must be his prey. Or else he is forced into the vicinity of human settlements, which lions find extremely distasteful.

There he often develops amazing cunning in hunting cows, other domestic animals, and sometimes even men. So-called man-eaters are usually exiles who attack human beings in desperation.

The fate of the lioness is quite different. Three and a half months after the honeymoon—by this time the lion has long since begun courting another female—she senses that she is about to give birth and once again isolates herself from the pride. But before she does so she finds another female who will assist in her confinement. Her choice falls either upon a lioness who is already too old to have offspring of her own, or upon a young 'miss', usually a daughter of her last litter. This assistance to the parturient is so common among lions that even the Masai tribesmen are aware of it. They call the helping lioness 'auntie'. Her task is to protect the mother and the newborn cubs from other predators, and to provide them with food.

As soon as the mother lioness has regained her strength, she goes back to hunting with her pride. At first she hides the cubs in crevices in cliffs, later in thickets, as Dr. Rudolf Schenkel[2, 3] of the Basel Zoo has observed in Nairobi National Park. Norman Carr once saw a lioness dragging a killed sable antelope weighing three hundred pounds almost a mile to her cubs—a tremendous feat of strength.

A curious ceremonial takes place when the cubs are three months old. Their mother leads them to the pride for the first time and introduces them to all the 'adults'. After this they are accepted into the band. If this rite were not observed, the cubs would probably be killed as strangers by unaware members of the pride. For the friendly or hostile behavior of one lion to another depends solely on whether or not he knows him personally as a member of his own pride.

Although the male has long ago turned his affections to another, the mother remains submissive to the father of her cubs. The result is that the father becomes a particularly good friend of his own children. And friendships among lions are stronger than death. If a mother dies, other mothers raise the cubs with as much devotion as their own. This kind of magnanimity, as rare in the realm of animals as it is among men, has been observed by Professor Bernhard Grzimek[4] in the Serengeti Plain.

The mother begins training the cubs to hunt when they are five months old. At first she shows them how to use their claws to skin the prey, and how to let full intestines slide through their teeth to squeeze out the contents—for lions are dainty eaters.

The cubs' first 'prey' is the tassel at the end of the mother's twitching tail. She uses it to pose more and more difficult problems in catching hold. Later the cubs also train alone: they chase butterflies and locusts in the grass and play at attacking their brothers and sisters. In the ensuing rough-and-tumble the cubs must constantly give the friendship signals, lest the game suddenly degenerate into real combat.

At the age of one and a half the adolescents are ripe for advanced schooling in actual hunting. Two or more mothers join their band of cubs and scout for game. As soon as the mothers begin stalking, the cubs try to imitate their every movement. At first, of course, they look extremely clumsy. But they soon learn to move far apart and form a kind of skirmish line. Each must work his way forward so skillfully that he will not be observed by the quarry.

Such hunts can go on for many hours of steadily mounting tension. Norman Carr describes what happened when at the very end of such a protracted hunt a half-grown cub spoiled the whole show by a single incautious movement. As the quarry raced away, the mothers rose, shook off their disappointment, and without punishing the guilty cub, without so much as giving a sign of displeasure, set off with infinite patience in search of fresh game.

Such behavior represents some fairly advanced pedagogy. In pursuit of a goal, human beings too can be trained to patience and perseverance only by the use of patience and perseverance. For a predator, such qualities are vital to survival. If hunting lions behaved like anthropoid apes, whose concentration span is limited to a few minutes, they would long ago have been starved into extinction. Let us remember this, for it is important for our later conclusions: for the life of the predator, patience and perseverance are essential.

Sometimes tragic family disputes arise between 'juveniles' and fathers. For fear of new rivals the old males frequently drive away all their male offspring. The youths form into bands together with fellow victims from neighboring prides. These young males have already been trained in hunting, but they lack experience. For them the most difficult period in a lion's life now begins. Reduced to skin and bone from starvation, constantly in flight from the prides that dominate the surrounding territories, most of these exiled young males die miserably.

Fights with established 'property owners' are a far cry from the tournaments among the members of the same pride. In both cases, to be sure, the conflict begins with some lusty roaring. Usually the entire pride assembles when it realizes that strangers have entered the territory. In 'chorus' it tries to howl the intruders away. But if that has no effect, bloody wounds are inflicted with claws and teeth.

Sometimes a young male in sheer desperation will simply refuse to be driven on any more. In spite of many defeats and wounds he returns to the territory again and again. Then, one day, the lions of the pride put an end to the dissenter. Without any further ado, they kill him. In this situation, and only in this one, do lions kill members of their own species. Once they have done so, they usually eat him as well. But it must be emphasized that these cannibalistic lions never kill a member of their species with the deliberate intention of eating him. Possibly the corpse proves to be a temptation too great to resist.

Carl Zuckmayer would have had to forfeit a good line if he had consulted behavioral scientists before writing his play, The Cold Light. For he has a nuclear physicist say: 'The only animal on earth who kills his own kind is man.' Murder exists in the animal world, and sometimes of a very ruthless kind.

However, we do not intend to moralize here on the matter. It is much more important to know what the situations are in which animals kill their own kind, and why they do so. Who knows, they may be responding to the same underlying forces that transform men into inhuman monsters.

The lion is undoubtedly an animal with a highly aggressive disposition. But within his clan he controls his bloodthirstiness so effectively that he can be called a friendly, even a tender-hearted beast. What is it that inhibits his aggressiveness when in the presence of fellows he knows well?

Scientists have observed gestures of placation, peacemaking, and humility in many species of animals.[5] When animals of the same species, wolves for example, fight with one another, the loser can prevent the victor's fatal bite by offering the most vulnerable part of his body, his throat. As if spellbound, the victor pauses; he is 'automatically' incapable of closing his teeth on the unprotected throat. That is a mode of behavior firmly fixed in the structure of his instincts.

Lions,[6] like human beings, unfortunately do not have at their disposal this astonishing and effective instinctual inhibition. Moreover, wolves do not depend solely upon this gesture for inhibiting the killer instinct. Their behavior leads us somewhat closer to the complex of problems with which mankind must learn to deal.

Just like a wolf, with this gesture of humility, offering his unprotected throat, a thoroughbred dog checks the attack of a superior member of his species.

The wolf Lobo and the wolf bitch Anastasia were the lowest in rank in the large wolf pack of Chicago's Brookfield Zoo. In the free wolf packs of northern Canada and Alaska it may happen that lowest-ranking animals, who cannot fit into the pack or are regarded as 'unsympathetic' by other members of the pack, may be driven away or even killed.

It might well have come to that point in the two-acre enclosure at the Chicago zoo if the wolves in question had not fundamentally changed their behavior at the last moment, and thus saved their membership in the pack.

Lobo managed it with a rather cheap trick, as zoologist George B. Rabb[7] observed. Whenever another wolf wanted to jostle him—for he was regarded as the whipping boy of the pack—he began limping piteously, as if he had broken a leg. But as soon as the bully was gone, Lobo trotted smartly away. The amazing thing is that the same pretense worked every time.

The wolf bitch Anastasia also realized that she was on the point of being expelled, and took far more skillful measures than Lobo. From one day to the next this outsider began making herself useful to three bitches higher than herself in ranking. She acted as nursemaid, cared for the others' litters, played with the cubs, and made sure that they did not drift away from the pack. From the moment she launched on this program, Anastasia was once more respected as a member of the social group and no wolf tried to drive her away or snap at her.

Thus the aggressiveness of wolves can be turned to friendliness by either pity or proofs of friendship. Aside from an instinctual mechanism that inhibits aggression, there exists a social cohesion which likewise sets up a clear inhibition of aggressiveness toward acknowledged members of the pack. This second phenomenon is also found among lions, among hyenas—and among men.

The Horde also Has Its Hunting Laws

From seventy throats ugly howls and infernal laughter break the darkness of an African night. It is the battle cry of two packs of hyenas engaged in a life-and-death struggle. Next day lions can be seen feeding on the carcasses of the dead hyenas.

This incident has been described by Dr. Hans Kruuk,[8] the Dutch behavioral scientist who spent more than a year at the Serengeti Research Institute in Tanzania studying these animals which hitherto have been so wrongly regarded as 'repulsive, stinking, and cowardly carrion-eaters'. His observations in the Ngorongoro Crater in 1965 and

1966 led to a completely revised picture of the defamed beasts. The hyena is a gregarious animal that hunts at night in well-organized packs which can wreak havoc even on large and healthy hoofed animals.

Hyenas are at most carrion-eaters by day, and apparently only when they have been unsuccessful in their nocturnal hunts. Only at such times, too, do they prey on newborn gazelles and gnus. Generally they sleep by day, each hyena by itself in a 'private' den, burrow, or thicket. They emerge at nightfall and always assemble at the same meeting place—usually a sizable common cave. Their gathering is accompanied by interminable ceremonial greetings until the full pack of some twenty head is assembled.

Animal greeting ceremonies are in a sense comparable to human gestures of friendliness. The hyena is by nature highly aggressive. When they meet their own kind, whom they know to be a member of the pack, the animals must neutralize their instinctive aggression by a ritualistic show of peace. They have a kind of gesture language which they use to express the fact that their approach is not with any hostile intent.

As soon as the whole pack is assembled, it trots off in close formation. Dr. Kruuk had learned how to follow the pack in a jeep on moonlit nights, and in this way witnessed about a hundred hunting expeditions.

Curiously enough, the hyenas do not fall upon the first zebra or gnu they come across. Even when these animals are grazing by the hundreds in the pale, moonlit plain, the night hunters seem to take not the slightest interest in them. As if they had previously determined on a particular hunting territory, they lope steadily ahead for several miles, refusing to be distracted. While running they constantly leave scent marks in the grass, presumably to inform other hyena packs of their presence.

In the hunting territory, their behavior changes abruptly. They raise their bushy tails high, and their noses sniff the ground. Then the hyenas approach within some five yards of the zebras.

Zebras live in large herds of polygamous families consisting of a stallion, from two to seven mares, and foals. At first the zebra family threatened by the hyena pack stands rigid, but extremely alert in the nocturnal darkness, watching the approaching enemies. Suddenly a stallion rushes forward in a bold counterattack. He flails out at the hyenas with his hoofs and snaps and bites, while his family flee in the opposite direction.

In semicircular formation, the hyenas take up the pursuit. But the zebra stallion constantly intervenes, galloping to the left and right, back and forth between his family and the pursuers. Since it is fatal for the hard-pressed zebras to lose touch with one another and with their stallion, they run at only a moderate pace. Sometimes they even pause until the stallion has caught up with them again.

Meanwhile each hyena follows its own quarry, trying to get in a bite at the leg or stomach of the nearest mare or foal (but not the raging stallion). When a zebra has been injured by several bites, it lags behind. Instantly all the hyenas observe this and close in upon this one victim. It stands helpless, without offering resistance, and soon falls to the ground and dies within a few minutes. Half an hour later not a bone or scrap of skin remains to bear witness to the nocturnal tragedy.

In thirty-three such hunts that Dr. Kruuk accompanied, however, the hyenas brought down their prey in only six cases. Twenty-seven times the stallion succeeded in driving off the pursuers. His courage, skill, and endurance are the decisive factors in saving the lives of his family.

In the course of his observations the Dutch zoologist made tape recordings of the howls the nocturnal hunters utter when they are tearing their prey to pieces. Replaying this at very high volume, he lured nearly sixty hyenas to his campsite. He stunned them with an anaesthetic gun and tagged them. This enabled him to follow the career of many individual hyenas for a year.

He discovered that the hyenas do not decide freely where they are to hunt. The packs must respect territorial boundaries. But there is a novel and unusual side to this principle: not every pack has a territory of its own. Instead, there are friendship agreements with neighboring packs that tolerate one another. Dr. Kruuk called these superordinate social organizations, embracing five or six packs, 'clans'.

Interesting social rules govern the relationships within a clan. If, for example, a pack has killed more game than it can consume at one time, it places perfumed lumps of excrement around the cadaver. Other packs, and solitary hyenas of the same clan, respect this marked property. They do not eat the carrion, however great their hunger.

If lions find the carrion, on the other hand, the rightful owners are out of luck. It flouts all we have been taught to think about the characters of the two animals, but the latest research seems to indicate that lions eat the carrion of animals killed by hyenas far more often than the other way around!

Disputes among hyenas of the same clan are generally settled by bloodless rituals of threat and appeasement. Since these animals are equipped with jaws capable of cracking the bones of elephants and rhinoceroses, a serious fight would invariably have fatal consequences. And because animals almost always live in accordance with the principle of incurring the least risk, they renounce the use of force within their society. In their case, however, it is not so much personal acquaintanceship that quells aggression, as among lions, but membership in the clan.

From clan to clan hyenas behave no better than men from nation to nation. In case of border violations, all inhibitions vanish and the most savage fights ensue.

In the Ngorongoro, clan boundaries tend to be crossed in the excitement of hunting zebra or gnu. In such cases, the prey belongs not to the hunters and killers, but to the clan in whose territory the game is brought down. If the kill takes place just barely beyond the border, and if the owners of the territory have not yet noticed, the hyenas—like human hunters in similar situations who have inadvertently shot their game across a state line, where the hunting season is not yet open—will try to smuggle their prey as quietly and quickly as possible back into their own territory.

The deeper the pursuit is continued into enemy country, the more critical the situation becomes. The hyenas of the alien clan are on the watch, and as soon as they hear their boundary violated they rush to the spot in order to assert their rights. What happens then depends partly on the relative numbers of the two hostile hordes. But other factors are more decisive: the self-assurance of the hunters and the depth of penetration into the enemy territory.

With increasing distance from their own familiar territory, the uncertainty and anxiety of the intruders increase. The result is that twenty or more strangers may be driven to flight by only three owners of the territory. But if in both packs of hyenas, hunger, depth of penetration, and strength of numbers bring the aggressive impulses to approximately the same pitch, a bloody war is inevitable, and ends with a battlefield strewn with corpses.

Do Rat Nations Wage Wars?

We have recently been hearing a good deal about a new villain in the animal kingdom: the rat. The oft-belabored comparison between totalitarian societies and the deterministic ant and termite states has proved to be oversimplified, untenable in many of its details. Recently, however, zoologists[9] have been calling the rodent, which in any case has long been held in horror by many persons, a prime example of misguided aggressive behavior. In both rat and man such behavior leads to the collective struggle of one community against another, and thus to senseless self-destruction.

The Glasgow zoologist S. A. Barnett[10] has investigated aspects of the social life and the fighting behavior of rats that shatter many an old legend about the unbidden guests in our cellars and attics.

The brown or Norway rat, so the story goes, first conquered Europe in the eighteenth century, defeating the smaller black rat. It is said that in 1727 an immense army of migratory brown rats swam across the Volga near Astrakhan and invaded the countries of the West like the Huns. As we have seen, lemmings march in armies of millions and cover as much as 150 miles. Rats do not behave this way, however. Their 'armies' are always groups in flight, not aggressive bands. Moreover, they run in crowds only for short distances, then scatter again as soon as the danger is over.

The 'invading army' which crossed the Volga at that time was probably fleeing from an earthquake which had just shaken the

Kirghiz Steppe. Rats also leave a sinking ship in great columns. This proverbial phenomenon has nothing to do with mystical prophetic powers, however. Being by origin burrow dwellers, the rodents live in the lowest areas of the ship, in the bilge, which is almost inaccessible to the sailors. They thus become aware of the water entering through leaks sooner than the crew. When their nesting places are flooded, they must perforce flee the ship. Their cries of alarm summon the others of their kind from the holds, and the result is a panicky exodus.

We have no idea what methods the Pied Piper of Hamelin employed. If the story has any kernel of truth at all, he may have managed to flood the cellars or frighten the rats in some other way, thus initiating a wild flight. The piping was certainly no more than an accessory bit of quackery.

In short, armies of rats marching into enemy country belong to the realm of fable. The brown rat which originally lived only in northern China has conquered the world in a different fashion: by a slow, tenacious house-to-house, village-to-village and city-to-city struggle it has gradually displaced the black rat.

Legend Number two: Hitherto scientists believed that when the archenemies met, the brown rat, weighing about a pound, tore to pieces the black rat, which is only about three-fifths as heavy. What is the truth of the matter?

In exciting experiments at the University of Glasgow, Barnett[10] checked this theory. In two adjacent rooms in his laboratory he raised two groups of the hostile species. One day he opened the connecting door. The large brown rats immediately entered the foreign territory and drove the owners from their nests to the farthest corner of their room. Within a few days thirteen of the nineteen black rats were dead. But they did not die from injuries or bites, nor from hunger. They showed no internal hemorrhages and no signs of infection. In fact no physical cause of death at all could be discerned. Barnett records feeling a distinct sense of the uncanny in the face of all this.

It is a cruel experiment to pit animals in a life-and-death struggle with one another. But Konrad Lorenz has remarked that 'the be-havioral scientist, haunted as he is by the destruction threatening humanity, must see in the swollen body of a dead rat his natural image'. Barnett was mindful of this prescript. He therefore threw some rats together, filming the confrontation, in order to analyze the sometimes extremely rapid developments among the hostile parties.

The scientist placed a strong brown rat in the territory of a strange tribe of brown rats. At first nothing happened, for the rats of a tribe do not recognize each other by appearance, but by smell; each tribe has its characteristic 'scent uniform'.

Suddenly a native rat caught wind of the intruder. It sniffed him, its hair standing up on end and its teeth chattering excitedly. Then it defecated and urinated, turned its flank to the enemy, crooked its back in the typical threatening gesture of a rat, and circled the stranger with affected-looking small steps, its legs stiff. The stranger meanwhile crouched petrified.

Then the rat defending its territory began a kind of Indian dance around the stake. With incredibly rapid prancing movements of its forelegs it darted directly at the interloper, who made no attempt to defend himself, leaped wildly around him, and nipped him once or twice in the tail, leg, or ear. The 'round' ended within a few seconds, but further attacks followed quickly. After the sixth absolutely bloodless repetition, the 'home' rat made a pause and left the scene for a while.

This did not mean that the war dance had tired the attacker. On the contrary, after his 'act of heroism' he promptly copulated with a female. But the victim, which had not stirred throughout the performance, was breathing rapidly and irregularly and showed all the signs of total exhaustion.

A stranger rat that is exposed to such sham attacks several times a day dies within a few days. In one extreme case Barnett observed that death ensued after ninety minutes. But in no case was a bite, a wound, or any other injury the cause of death. The animals which were strong and healthy at the beginning of this treatment slowly faded away. The English scholar could not even discover any physical indications of death by shock.

The mystery of these deaths was first solved in the course of 1967. At

King's College in London Dr. Gillian Sewell[11] discovered the ultrasonic language of rats. The aggressor screams at his victim in a series of ultrasonic bursts inaudible to us. Each burst lasts only thirty to sixty milliseconds, while the rigid victim, frozen in terror, gives vent to a kind of gasping: ultrasonic cries that last for as much as seven thousand milliseconds. Presumably the intensive ultrasonic aggression literally numbs the stranger's nerves and eventually destroys them.

It is astonishing that the attacked rat makes no attempt at resistance. This is strange behavior in an otherwise fierce animal which when cornered by dogs and men will utter a piercing battle cry and spring at their throats. Apparently stiff defenselessness is a kind of appeasement gesture and is the only way to prevent their own kind from actually launching bloody assaults.

In the open, such behavior has great survival value, for there the conflict takes an entirely different course from what happens in the closed arena of the laboratory. After a series of 'war dances' the attacked stranger has an opportunity to run away unharmed. He therefore does not die. Thus rats in general wage a kind of cold war.

In fact the brown rats that advanced from Asia to Europe did not exterminate the already settled black rats. They merely drove them from the cellars to the attics, where the brown rats do not feel at ease. Thus both species can live one above the other in a house—with the stairs and living rooms as a no-rat's-land between their respective territories.

Can rats then be said to fight collectively, tribe against tribe? What actually is the social structure of such a tribe which, for example, defends the cellars of a farm against 'foreigners' from the surrounding farms?

Among most other animals that live in groups, such as wolves, apes, and hens, the social structure is based on a linear ladder. Rats, however, regulate their community life by a kind of three-class system. Professor Barnett has discovered that there are alpha, beta, and gamma rats. Within each class an absolute equality of rights seems to exist.

The alpha rats represent the aristocracy, as it were. All other males, and the females as well, yield to them with a gesture of humility. If two alphas encounter each other, they make brief threatening gestures and then part without excitement. They are the ones who drive off any intruders.

The beta rats look just as large and healthy as the others, but they behave submissively not only toward alphas of their own tribes, but toward strangers as well. When a rat assumes beta status, the result is almost total extinction of the aggressive instinct.

But who or what decides the class position of each individual. That is one of the most interesting questions in rat research, for evidently no 'classification battles' take place among these rodents. The first hints of a solution were obtained by Dr. Hilary Oldfield-Box[12] at the Psychological Laboratory of Sheffield University. Her findings caused an instant stir in scientific circles.

Dr. Oldfield-Box trained a number of adult brown rats to earn their living as 'factory workers'. The animals had simply to press a lever down, but their pay was scanty: each time they received only a single grain of wheat.

After the rats were trained, the scientist placed three of them together in a cage which contained only a single feeding lever. At first all three rats worked it alternately and on a fairly equal basis. But soon specialization developed: one animal continually stepped on the lever and was therefore working very hard, while the two others ate most of the food from under his nose. In over a hundred similar experiments the same pattern developed: the group divided up into one worker and two eaters!

Curiously enough, the class division took place without conflict every time, even without threats or attempts at self-aggrandizement. It seemed to happen of its own accord. What is it that decides the social position of each individual?

Dr. Oldfield-Box placed heavy rats with light ones, large ones with small ones, old ones with young ones, males with females. In no case could she predict which ones would become the workers and which the eaters. The classical categories of social theory, ranking, dominance, submission, and cooperation, all failed to fit the case.

When new groups were formed, consisting only of proved workers or only of eaters, a new hierarchy developed. Of three former workers, two would become eaters while the third slaved for them. But as soon as they were returned to their original group they resumed their old position as workers. Even when Dr. Oldfield-Box separated one group of three, let the animals 'forget' their experience for weeks, and re-assembled them again in the old way only after a long time, each rat promptly assumed its former position. Therefore the distribution could not be a matter of chance.

Obviously there is some sort of social order which is not marked by suppression of the weaker by the stronger, nor by superiority and unscrupulous exploitation—and this social order is found among rats, of all animals! It would be fascinating for us human beings to discover the hidden social forces at work in these relationships. Unfortunately the British scientist has not yet been able to solve the mystery. Nevertheless, there are indications that all those animals that at the very start of the experiments, during their training in individual cages, showed themselves most active and intelligent are predestined—to be workers.

This conclusion would agree with an observation of rat tribes under conditions of freedom. Among these there are, as has been mentioned, no rankings in the ordinary sense within the various classes. On the contrary, a rat that seems to us strong and well nourished always allows weaklings, females, and young rats to take precedence at feeding. But it has not yet been determined whether Dr. Oldfield-Box's worker rats are identical with S. A. Barnett's alpha animals. Barnett, as we have noted, also speaks of gamma rats. These have rough, dirty coats and are of sluggish disposition.

The fighting force of a rat tribe consists of only the relatively few alpha rats. But even among these not a single case of cooperation was observed in Glasgow—neither in the search for food nor in fighting. The scent of an intruder naturally stirs a certain excitement in the territory, and every alpha rat smells with special care every other rat it meets, to determine whether it is a member of the tribe. But the fighters act only on their own individual initiative. Nor are there any 'generals' in command of an army of rats.

The only cases of rats actually biting each other to death occurred when laboratory rats were put in an enclosure with wild members of their own species. The laboratory rats and their forefathers had been living for decades in an unnatural environment. They had lost the ability to communicate with others of their kind and to respond to social signals. Ignorant of the gestures of humility and appeasement, and of the signs of friendliness, they behaved wrongly when they were attacked. Their deaths were only the result of a kind of misunderstanding.

It thus becomes clear that the animal analogue of bellicose man, the rat, in fact offers no resemblances at all that can help illuminate the ominous world situation.

6. The Attraction of Evil

Beyond Eden

The two rats which Professor Roger Ulrich[1, 2, 3] had locked up together in a cage were good friends. But when the American psychologist administered an electric shock to the foot of one of the two animals, the peaceful picture changed abruptly. Under the influence of pain the rat flew at his companion. The longer the pain lasted, the more ferociously he attacked his innocent cagemate.

This result led the Western Michigan University scientist to an interesting conclusion: 'In animals and probably in man as well aggressive behavior is an immediate and natural consequence of pain.' Konrad Lorenz's theory of aggression (*see* Chapter 3, note 33) can be extended and further refined. We may say that pain lowers the quantum needed to trigger the instinct of aggression.

Biologically, the apparently absurd reaction observed in the above experiment has its purpose. An animal that is attacked, and therefore feels pain, has to be impelled to take defensive measures. In general, therefore, the aggressive response to pain is meaningful behavior. But the experiment shows that the impulse, though biologically reasonable, does not operate logically in terms of human reason. That is why psychologists find so much that is incomprehensible in belligerent behavior.

The irrationality of belligerence emerges most clearly when the counteraggression stimulated by physical or psychic pain has no meaningful outlet. The biological belligerence has been called into play; what is more, it has been heated to high pressure and must somehow blow off steam. Unfortunately this is as far as the physiology of the instincts goes; what happens after that is a matter of complete indifference to nature. The shocked rat felt a pain against which it had no defense. It therefore picked on a scapegoat.

Or let us take another example, not an artificial laboratory case. A chicken that is 'punished', that is pecked at, by a stronger member of the flock cannot discharge its aggression, which erupts 'automatically', against its tormentor. That would have dire consequences. Instead it promptly chooses a weaker, though completely innocent fellow and pecks away till the feathers fly.

This 'pecking order' effect appears in the social life of both animals and men. It has nothing to do with rationality and morality, but rather arises from the very structure of the instincts. It, too, represents one of the animal roots of human nature. In the light of this principle, we can better understand why an ill-natured, harsh boss creates an atmosphere in his shop in which everyone down to the lowest levels is hostile to everyone else. Moreover, the 'illogical' consequences of the boss's temperament is that work is slowed down instead of being stimulated.

An instructive marginal phenomenon emerged from Roger Ulrich's experiments. Given electric shocks of uniform intensity, the fury of the rats toward their innocent cagemates increased in inverse proportion to the size of the cage. The smaller the cage, the closer together the two animals were thrown, the more angry the shocked rats were. 'You can kill a man not only with a club, but also with too small a room.' This was a proverb popular in Germany after the war, among refugees and bombed-out people. Evidently in this respect human beings do respond like rats. Witness the everlasting bad blood between families who sublet rooms in their apartments and their tenants. The tenants are not people of such offensive character. Rather, the disputes arise from the conjunction of disposition, displacement, and the spatial factor. Sensible people should regard such rows as a sort of natural phenomenon, like rain and thunderstorms. Such feelings must be understood as a direct result of the housing shortage, and everyone concerned must refrain from making the situation worse by moral judgment and reprisals.

But is reason of any avail in such cases?

Professor Ulrich's dramatic discovery seems to suggest that it is not. In his experimental box he installed a lever which the rat had simply to press down in order to stop the electric shock. Animals kept by themselves quickly learned the routine. They also learned to sit close to the lever, so that they would be able to activate it quickly as soon as the cage was electrified.

But when an animal who had perfectly mastered this technique was brought together with another rat, the results were grotesque indeed. Although the shocked animal knew the quickest way to eliminate the pain, in most cases it did not act rationally, but instead attacked its cage-mate. Apparently the rat felt a greater urge to punish a whipping boy than to switch off the source of the pain—which continued undiminished during the fight.

This experiment rather dramatically illustrates how aggression kindled by pain tends to block rational and learned behavior and lead even a trained animal—and probably a human being—to try to put the blame on another.

Such is the tragic interrelationship of instinct and reason. It is no less tragic that in their efforts to deal intellectually with the phenomenon of evil, men have considered it in purely moral terms. Despite the convincing picture drawn by Konrad Lorenz in 1963 of the other, the instinctual side, of aggression, there is still a widespread failure to recognize that so-called evil may involve a biologically fixed component. Hence we have all the more reason to look at certain new researches which are highly pertinent.

Professor Walter C. Rothenbuhler[4] of Ohio State University may well be called the Gregor Mendel of behavioral research. In 1964 he provided proof of a hypothesis on which leading biologists throughout the world had been divided for decades, some passionately affirming it, some vigorously denying it. The hypothesis is that not only physical shapes, but single elements in the behavior of animals, are firmly fixed in given genes of their hereditary substance.

In experimenting with honeybees the American entomologist observed that he had two genetic lines of bees which differed solely in two aspects of behavior: aggressiveness and cleanliness. The workers of Line A were very gentle. In numerous tests they stung an assistant only once. But at the same time these peaceful bees were very unclean. When larvae died inside their comb, either of disease or artificial injection of poisons, they unconcernedly let the bodies decay.

Working under similar conditions with the aggressive bees of Line B the scientists received no less than 150 stings. It was also noteworthy that these bees would hurry to remove every dead larva from the hive.

Rothenbuhler had the brilliant idea of crossing these two types of bees. The hybrids of the F_1 generation behaved 'unhygienically' without exception. F_1 hybrid queens were then crossed back into the parental lines. The twenty-eight hive populations that resulted from these crossings revealed astonishing characteristics.

In seven of the hives the workers at once opened the cells affected by foul brood and cleaned them. In seven other hives they opened the cells, but always let the dead larvae remain in them. In another seven hives the bees removed the dead larvae, but only if the experimenter had beforehand attended to the job of opening the cells. In the last seven the workers did not open the cells, nor did they remove the dead larvae even if the cells were opened for them.

Gregor Mendel had not provided a more precise demonstration of inherited characteristics in his experiments with the color of pea blossoms!

This unequivocal result permits only one conclusion: in the genetic material of the bee chromosomes there is a recessive gene for the opening of cells with dead larvae, and another recessive gene for carrying away the dead larvae. In order for the hive to be cleansed of disease-carrying dead brood, a worker bee must have received each of these two genes from both parents, that is, from the queen and the drone. For both modes of behavior they must be what biologists call homozygous.

The full implications of this discovery cannot yet be foreseen, but at least two aspects have become clear. The first is: unfortunately, it is impossible to breed gentle bees that hardly ever sting, for in these insects gentleness is linked with a congenital inability to cleanse the hive of foci of disease.

Secondly, Rothenbuhler's experiment reveals fully the complicated stratification of instinct. Laymen still speak in rough generalizations of the instincts of feeding and copulation. But the late Professor Erich von Holst,[5] former Director of the Max Planck Institute for Behavioral Physiology at Seewiesen, pointed out a number of years ago that the so-called 'major' instincts are genetically compounded of numerous 'minor' instincts.

In the dog, for example, sniffing, tracking, running, hunting, and shaking the prey to death represent completely independent inherited instincts. But they are so harmoniously interrelated that in conjunction they enable the animal to satisfy its hunger. It now appears, moreover, that the instincts can be even further subdivided. With the bee, at any rate, we can no longer speak of an instinct for keeping the hive clean; but rather of an instinct for each of the processes involved: an instinct for opening the cells containing dead brood, and another instinct for carrying away the dead larvae. Only the two instincts working together merge into an act that is meaningful to the survival of the hive.

And, as it happens, aggressiveness is found in one of these genes. This is not to suggest that in man as well aggressiveness and cleanliness are located in the same gene. To be sure cleanliness is carried to extremes in military barracks throughout the world. But any conclusion by analogy from bee to man is not scientifically tenable. What is important is to discover that the instinct of aggression is genetically fixed, not only in the bee but even in man. That has been confirmed by two recent research papers.

In 1967 a British team studied the inmates of prisons at Rampton and Broadmoor to determine the composition of the chromosomes in the cell nuclei. Since there is considerable disparity in aggressiveness between men and women, Dr. M. D. Casey (Sheffield University) and Dr. W. H. Court Brown (Research Center for Radiation Effects in Edinburgh) examined the sex chromosomes with particular closeness.

Of the twenty-three pairs of chromosomes possessed by everyone, twenty-two are more or less the same for both men and women. There is, however, an additional pair which in women consists of two X chromosomes, in men of an X and Y chromosome. As a result of errors in cell division during maturation, however, human beings may sometimes have an extra X or Y chromosome in all their body cells. The results are as follows:

So-called XXY men have distinctly feminine traits. They have one female X chromosome too many. But there are also XYY types. These are supermen, so to speak. Most of them have come into conflict with the law in their early years for a wide variety of acts of violence. Among inmates of prisons they were found thirty times more frequently than in the normal population.

Following a totally different procedure, the second team of scientists arrived at the same conclusion. At the Institute for Arthritis and Metabolic Diseases in Bethesda, Maryland, J. A. Seegmiller, F. M. Rosenbloom, and W. N. Kelley[6] investigated male patients who displayed obsessional aggressive behavior, and sought to determine whether they deviated biochemically from normal. They found that a certain enzyme was almost completely lacking in such persons. The reason for the lack of this one enzyme may well be due to a failure or absence of the corresponding gene in the chromosomes.

The authors conclude: 'The association of a specific enzyme with a neurological disease, mental retardation, and a compulsive aggressive behavior may serve to reorient our fundamental approach to other behavioral disorders.'

Interestingly enough, the aggressive instinct can take radically different forms in whole groups of human beings. The Ute Indians who live on reservations in Utah and Colorado are a quick-tempered quarrelsome people, while the 'moon-children' on the palm-shaded coral reefs of the San Blas Islands, off the Caribbean coast of Panama, display a beatific peacefulness and moderation.

The prairie Indians of North America[7] must have developed intense aggressiveness during the several centuries when they were almost constantly on the warpath. Sexual selection led relatively quickly to corresponding biological changes in their heredity.

Today these congenitally aggressive Indians must live placidly on their reservations. They can no longer discharge their aggression in the old fashion. The consequence is that their blocked aggressiveness seeks other outlets, for example at the wheel of an automobile. To be sure, this is happening nowadays throughout the world. But among the Indians the number of automobile accidents far exceeds the frequency of accidents in any other group of drivers. A touch of ill humor, a moment's irritation at some other driver—and they promptly crash into their 'opponent'.

The behavior of the moon-children is as different as could be. These Indians of the Central American tribe known as the Cunas have become albinos as a result of mutation; they are white-skinned and white-haired. Albinism is a common phenomenon, especially among rabbits, rats, and mice. It has also been observed that the same mutation that causes the white coloration likewise changes the character of these animals. Albinic animals are far less inclined to bite than the normally colored members of their species.

It is curious and rather moving to find that the moon-children of the San Blas Islands follow the same rule. Their white faces with snow-white hair and deep blue eyes have an unearthly beauty. The anthropologist Clyde Edgar Keeler[8] of the University of Georgia has observed that their gait and movements are very deliberate. They speak softly and scarcely ever act vigorously. In quarrels among themselves or with their brown-skinned fellow tribesmen they have little perseverance; they soon give in and avoid physical clashes. This is not only the proverbially wiser course, for their intelligence is far above the average of their pigmented fellows.

But this is only one aspect of the missing aggressive instinct. Unfortunately there is a regrettable side to it. The moon-children never laugh and only very rarely give a faint smile. They do not take part in the festivals and pranks of their brown fellow tribesmen any more than they share their depressed or anxious moods. They can endure with stoic equanimity a raging hurricane that seems about to sweep the entire island into the sea. In spite of their intelligence they never undertake things of any great importance. And above all they seem incapable of feeling more than a feeble flicker of physical love. Albino women rarely have children.

All this illuminates how great a part the aggressive instinct plays in basic human psychology, and in the life of all animals, for good as well as for ill. The instinct for conflict is not only the primal element in destructiveness, but simultaneously an indispensable component of the enterprising spirit, of the capacity for enthusiasm, of all creative impulses—and of love as well.

If in the interests of a peaceful world, not only without wars but without personal quarrels, we were to attempt to breed the aggressive instinct out of the human race—and biologically that should be possible—we would make ourselves into moon-children.

The problem of aggression therefore cannot be solved by eliminating the aggressive instinct. Our difficulty is that we must balance upon a very narrow ridge between two abysses. Insufficient aggression would rob life of its force; excessive aggression means the unleashing of wholesale destruction.

We cannot therefore take a wholly positive or wholly negative view of aggression. This relic of animal behavior, the instinct as such, is not evil in itself, but rather the biological basis for everything that man's intellectual failures make into actual evil. It is only the fusion of elements of animal behavior with typically human traits that gives rise to the original sin which banished us to the lands beyond Eden.

What Drives Man to Inhumanity?

As soon as the aggressor entered the pacifist's home he rushed full tilt at his innocent victim. But as soon as the latter made a first few clumsy attempts at self-defense, the attacker was pulled back out of the room by a line attached to his foot. Thus the defender thought he had won, and gained courage when the same events were repeated on the following day.

By the third day the lust for battle had developed in the pacifist to such a point that he pursued an innocent third party and gave him a trouncing. After five days his reeducation into a fierce fighter was complete. His easy victories had made this originally gentle soul so aggressive that henceforth he even attacked, kicked, rammed, and bit 'women and children'.

Granted, these were not men being trained in this school of belligerence; they were mice. But the aggressiveness, destructiveness, and cruelty of men have similar causes, according to the American zoologist Dr. John P. Scott,[9] head of the Jackson Laboratory in Bar Harbor. He argues from his experiences with animals that evil in either animals or men is not inherent. Only destructive training can make both aggressive. Forced to fight, an animal wins, and its very success stimulates its aggressiveness. Presumably this also applies to human beings, says Dr. Scott.

On the other hand, by positive instead of negative training techniques Dr. Scott was able to build a social group of peacefully oriented mice. One of his methods was as follows:

Since females never dare to attack a male, to induce peaceful behavior it suffices to have the animals grow up in pairs, without contact with 'wicked' neighbors. There is then no challenging behavior in the mouse's environment and not the slightest need for it to begin a quarrel. Consequently, no disputes take place. The peaceful atmosphere so reinforces the male's peaceableness that he remains amicable when he is later introduced into a larger society of males who have likewise grown up in the pacific company of females.

Since among human beings, in contrast to mice, males and females sometimes get along quite badly, this experiment cannot very well be applied as a patent medicine for human contentiousness. Nevertheless, the result spurred Dr. Scott to seek other methods.

Certainly it would be one of the most tremendous discoveries of the past two thousand years if the problem of human combativeness, which has persisted despite the labors of all reformers and idealists, could be solved as with mice by suitable methods of education.

Other animal experiments by the American behavioral scientist Professor Z. Y. Kuo[10] seem to further this hope.

Kuo raised kittens under different educational systems. Twenty kittens in the first group were allowed to have a mother with them and were often privileged to look on when she caught rats and ate them. It was not surprising that eighteen of these kittens began hunting rats as soon as they were physically strong enough.

Twenty other kittens were left to themselves, without any parental model. When this group reached adulthood, only nine would spring at a rat. With these nine, obviously, the passion for hunting was in their blood.

The great surprise came with the twenty parentless kittens of a third group who were given a baby rat for a playmate from infancy. As this group matured, only three could not control their awakening killer instinct, while fifteen of the cats lovingly licked the by now grown rat whom they knew personally, and even protected it against their savage brothers and sisters. In short, an animal Eden of sorts had been created.

But can aggression really be eliminated by presenting children with a false world in which there is no aggression? Scott and Kuo incline toward a theory of pedagogy that the American scientist J. Dollard[11] conceived as early as 1939 under the impact of the aggressiveness of Hitler's Germany. He developed the hypothesis that aggressive behavior is always a consequence of prohibitions and frustrations. Therefore a child must be allowed to grow up freely; he must not be denied things, not be scolded, and above all never whipped. Many families, especially in the United States, have raised children by this doctrine of 'permissiveness'.

Unfortunately, the results have not justified the theory. The tolerant parents raised little family tyrants who found a positively diabolic pleasure in destruction and in terrorizing others. In subsequent years, when other adults took a strong stand against their uninhibited behavior, these children came down with a variety of blocks and neuroses. All in all, the permissive approach proved to be a poor way to raise human beings.

The source of error can be clearly observed in Kuo's experiment. Even among the cats raised in close friendship with a rat, three of the animals (15 percent) subsequently wanted to kill and eat the friend of their kittenhood. Given the circumstances, their passion for hunting could not have been learned. Consequently, it was innate. The extent to which training could suppress the inherited aggressive instincts of the cats depended on how strong those instincts were originally.

Scott's and Kuo's experiments offer convincing proof that training can influence inherited behavior. The factor of education emerged so strongly as almost to blot out an inherited component of behavior. But the component continued to be present. And if it is not properly taken into account it can shatter the whole concept—as in the case of the human children raised by permissive methods, in whom the aggressive factor spontaneously came to the fore.

The question, then, of whether man is by nature good or evil or whether his experiences in our world make him the one or the other, has been wrongly posed. Everyone has within himself by birth a more or less powerful aggressive component, and everyone's aggressive behavior is more or less subject to alteration by education and other environmental influences. A person with a highly aggressive disposition can keep himself so thoroughly under the control of his intellect that he reacts more pacifically than a less aggressive person with a smaller quotient of self-discipline.

In situations of extreme psychic pressure, however, there is a point at which the elemental drives take over. All men reach this point sooner or later. Depending on natural endowment and moral force, this point is found at different degrees of stress. It often happened in prison camps that someone who was always preaching morality would one day be caught in some disgusting act and henceforth be ridiculed by blacker sheep. Both the preachers and the mockers were mistaken in their view of human nature. Unfortunately there is no such thing as absolute good on earth.

Nowadays it is impossible to overstress this two-headed aspect of aggressive behavior, this alternation between good and evil, between animal and angel. For since the publication of Konrad Lorenz's book *On Aggression* there is a vogue for dangerous simplifications. A century ago the zealots who called themselves Darwinists but lacked the detailed understanding of their master vulgarized the doctrine of 'struggle for life' into the concept of 'eat or be eaten', and made it a formula to explain all political and economic behavior. Today, similarly, people who incline to violence loudly proclaim that there is no such thing as evil. All we have known by that name is only part of man's natural aggression, and hence every disgraceful act is in accord with nature. On the other hand, it is difficult to believe that behavioral studies of mice, rats, cats, and other animals, brilliant though the experiments may be, can actually provide us with the means for solving the great problems of human aggressiveness. 'Anyone who experienced the ghastly events in the Third Reich, the ruthless annihilation of millions of sentient fellow men, the slaughter of two world wars, which all the talk of cultural bonds in no way prevented or reduced—anyone who has seen these things in his own lifetime may find such a phrase as "so-called evil"* sticking in his throat,' writes Eugen Gürster,[12] who takes a dim view of what can be accomplished by the application of behavioral psychology.

The Nazi death camps of the recent past are without precedent in the animal world. But we are nowhere nearer the heart of the problem when we condemn the particular horrors of Fascism. Since the days of Hitler there have been too many massacres in other parts of the world. In the partitioning of Pakistan and India in 1947–48, between one and two million people were killed. After Burma's declaration of independence the victims were scarcely fewer. The violence in Colombia has taken the lives of countless innocent persons. An uprising in the Sudan is estimated to have inflicted death on some two hundred thousand people. And the anti-Communist outbreak in Indonesia in 1967 was even more murderous. Nor were the victims necessarily Communists. In the universal frenzy that swept over the country many people denounced their personal enemies, rivals, or creditors, and set the maddened hordes against them. Where will we find a similar phenomenon in the animal world?

We need only imagine ourselves in the midst of a mob bent on killing, in a group of religious fanatics, or in the office of an SS branch, to realize at once the hopelessness of any attempt to check such manifestations of inhuman cruelty. We need only read Jean-François Steiner's book on the revolt in the Treblinka death camp.[13] It is a minute and shattering documentation of what human beings are capable

* The original German title of Konrad Lorenz's well-known book *On Aggression* can be translated as *On So-called Evil*.

of, 'masters' as well as victims. Faced with these monstrous horrors, we cannot help feeling skepticism about any experiments that propose to throw light on human evil by examination of animal behavior.

Nor can such inhumanities be dismissed on the grounds that 'the Germans' or 'the SS' were extremely aggressive by nature. The fact is that conspicuous sadists (who the whole world over have a tendency to become prison guards) were removed by the SS leadership wherever they drew attention to themselves. Grotesquely enough, the SS leaders insisted on 'decency and propriety' in their personnel. These 'decent and proper' underlings were then *trained* to commit atrocities. Excessive aggression is therefore not the problem. Rather, the problem is that such appallingly large numbers of normal people did not scruple to carry out the orders of a few morbid sadists at the head of the government.

How many or how few persons possess the moral force to refuse obedience to amoral authorities? What archaic elements of behavior drag us into irrational and inhuman acts? Can we gain any better understanding of these forces from the behavior of social animals?

In a series of seven tests the American psychologist Dr. Stanley Milgram[14, 15] attempted to discover whether his authority would suffice to persuade experimental subjects to inflict pain and even death on other human beings. He succeeded all too easily.

The psychologist brought in a total of 280 persons, randomly selected from the street, to test their capacity for cruelty. He did not talk about ideological principles or other 'higher' ideals, but merely offered a deliberately low fee and pretended that he was conducting an experiment in educational methods. The problem allegedly was to find out whether pupils would learn to solve certain problems better or worse under the impact of progressively harsher punishments. The teacher, he informed them, would have to ask a series of prescribed questions and punish each mistake with increasing harshness.

The scale of punishments was truly draconian. First Dr. Milgram, assisted by the newly recruited teacher, strapped the supposed pupil into an electric chair which stood in a room by itself. In the teacher's room stood the punishment apparatus: a row of thirty buttons with which the pupil could supposedly be given electric shocks ranging from 15 to

450 volts. In addition to the numerical indication of the voltage, each key had a descriptive label of its effects, ranging all the way from 'light shock' to 'danger: severe shock'. Of course no real electric charges were involved. The supposed pupil was in reality one of the scientist's assistants, and his pain was entirely simulated.

In the first series of experiments the two rooms were so far apart that the 'teacher' could not hear the 'pupil' screaming. Although everyone ought to know that a 220-volt shock from an electric stove can be fatal, almost all the recruits ran blithely through the whole scale of punishments up to 450 volts. In other words, abstract knowledge of possible consequences presents no barriers to acts that will induce these consequences. In the event of war, men will undoubtedly as unfalteringly press those 'buttons' that will start rockets with nuclear warheads on their way.

In subsequent series, therefore, the experimenter set out to show the 'teacher' the effects of his punishments, in order to see at what point these impressions would stay his hand.

The electric chair was now moved to an adjacent room. At first Dr. Milgram permitted only a few of the pupil's reactions to reach the teacher's room. When the shocks reached 300 volts, the victim hammered against the wall. At 315 volts the agonized knocking suddenly stopped, as if the pupil were dead. In spite of desperate noises from the pupils, no less than 66 percent of the 'teachers' pressed the 315-volt button. Only 34 percent refused to do so.

Disturbed by this result, Dr. Milgram devised a further series of experiments in which he broadened the scale of acoustic manifestations of pain. At 75 volts a tape recorder automatically played a slight whimper behind the wall. At 120 it played a sober statement that the electric shocks were inflicting serious pain. Thereafter the reactions rose to energetic protest and a demand that the experiment be stopped and the pupil released. From 180 volts on there were sounds of loud screams, pleading, and an insistent cry that the pain had become unendurable. Above that voltage nothing but tormented screams were heard.

Depressingly enough, this experiment showed little significant difference from the preceding one. In spite of the pleas and screams,

62·5 percent of all test subjects administered successively stronger shocks to their victims. Several of the 'teachers' seemed, however, to falter. They asked Dr. Milgram, who stood by throughout the experiments, whether they should not stop the test. In such cases he did not reply with a stern order, but simply made a matter-of-fact statement: 'You have no choice. You must continue.' That was all.

What happened thereafter may be called an example of the classic discrepancy between words and deeds. Full of moral scruples, the teachers declared that they didn't want to hurt the poor fellow in the next room, that they found all this very unpleasant, and so on. Nevertheless, bowing to the scientist's authority, they went on administering stronger and stronger shocks to the pupil. Only 37·5 percent refused to go on working for their 'employer'.

Would close personal contact and actual sight of the victim decisively shift the balance? Dr. Milgram hoped to achieve this by placing teacher and pupil in the same room, with only a foot and a half distance between them. His assistant developed remarkable histrionic gifts in his role of pupil, from jerking briefly at supposedly weak electric shocks all the way to making his whole body quiver, uttering unnerving screams, and writhing as if in his last agony. It made little difference. The victim complained of pains in the heart, but 40 percent of the teachers went on administering the shocks when they were ordered to do so. The victim pleaded to be released, but 40 percent of the test subjects stepped up the pain.

'Initially we had not imagined that such drastic procedures would be needed to evoke refusal to obey,' Dr. Milgram writes. 'With stunning regularity we saw good people submit to the demands of authority and carry out acts that were unfeeling and harsh. People who in daily life were decent and conscious of responsibility were seduced by the imposition of authority and by uncritical acceptance of the experimenter's definition of the situation into committing cruel acts.'

In forcing the subjects to commit a direct physical act of violence, their moral resistance might be aroused. Therefore it was explained that the pupil could feel the electric punishment only if he kept his hand in contact with a steel plate on the arm of the chair. From 150 volts on the pupil refused to do this. The scientist thereupon ordered the teacher to press the victim's hand against the plate by force.

The very first test subject followed instructions, applying force against the victim and administering electric shocks up to the highest degree! And no less than 30 percent of the subjects behaved with equal cruelty.

In the sixth series of experiments, Dr. Milgram tried to see what happened when the influence of authority was diminished. He gave his instructions over a telephone from another room, instead of supervising the experiment directly. At once many of the teachers began to cheat. They gave their victims weaker electric shocks than they reported over the telephone. But even under these conditions there were still 10 percent who carried out these inhuman commands punctiliously, 'only doing their duty'.

A final variation of the experiment is highly illuminating. Before they themselves took their turns as 'teachers', the test subjects were given the opportunity to watch a predecessor perform the brutal procedure. Even then 10 percent of the participants were not deterred from going through with the experiment, and inflicting a fresh round of punishments on the victims. The fact that in these circumstances 90 percent of the subjects refused to obey is no consolation, for 10 percent of the population are more than enough to tyrannize over the rest of their fellows.

'The author finds these results disturbing,' Dr. Milgram sums up, 'They suggest the possibility that we cannot expect human nature or, more specifically, the type of character produced by American society, to protect citizens from brutal and inhuman treatment on the orders of an evil authority. On the whole people do what they are told to do, disregarding the contents of their actions and feeling no conscientious limitations, so long as they see the order coming from a legitimate authority. When in this study an anonymous experimenter could successfully order adults to force a fifty-year-old man into the harness and to inflict painful electric shocks upon him in spite of his protests, we can only wonder what a government—with far greater authority and greater prestige—can order its subjects to do.'

In Germany we no longer wonder. We know—from the Nazi period. And the mass slaughters in many other parts of the world provide depressing corroboration in political practice of the laboratory results. History has made it all too clear that man's inhibitions against killing vanish with terrifying swiftness when he thinks that he is acting for the sake of a religious ideal, that he is performing a good deed, pleasing to God. In such cases human ferocity no longer encounters the slightest inner resistance.

The Old Testament illustrates this ambivalence of the human psyche in the story of God-fearing Abraham who was ready to sacrifice his own son at the command of God. 'In the symbolism of Abraham's sacrifice,' writes Irenäus Eibl-Eibesfeldt[16] in his commentary on Dr. Milgram's experiments, 'is undoubtedly to be found one of the greatest human problems. Obedience is an ethical principle, just as love of neighbor is. But when does it cease to be ethical? If both these ethical principles come into conflict with one another, obedience often proves to be the stronger, apparently on the basis of an inborn disposition whose roots probably go back to the ranking orders of our apelike ancestors.'

A group of social animals will generally find it advantageous to follow their strongest, and usually most intelligent, leader. But the leader must first acquire authority by fighting for rank, by special achievements, and above all by general acknowledgement of this authority (see pages 70 and 75).

The capacity and willingness to accept subordination is not innate in all animals. It is missing in nongregarious animals such as cats, turtles, and badgers. Dr. Eibl's tame badger always remained self-willed and could not be forced to do anything. If the zoologist tried to punish him for a misdeed by slapping him, the badger immediately turned against him. On the other hand a dog adjusts to his master and often joyfully subordinates himself. Like young ravens and most apes, he is by nature a creature obedient to authority.

'It follows, however, from these observations,' Dr. Eibl-Eibesfeldt concludes, 'that love of neighbor and an individual's moral feelings are often not enough to make him defy the contrary commands of strong authorities. In times of peace mankind accepts certain humanitarian standards. Were these to be ratified on an international basis and spelled out in detail (as, for example, the Hippocratic Oath for doctors), it would represent a significant humanitarian advance. The individual could then appeal to the abstract authority of a law against the commands of evil authorities, and would thereby find support. In his moral decisions he would no longer be standing alone against authority, but would have another authority as his ally.'

This seems a more cogent solution than others that have been proposed.

American veterinarians[17] at the Experimental Institute for Animal Diseases in Ames, Iowa, have hit on a way to transform vicious predators into peaceful, docile creatures by special feeding. Ferrets continually bit the scientists unpredictably and without provocation. But when they had been fed a special diet, they became as tame and docile as lapdogs. The diet which so profoundly 'improved their character' did not consist, as one might assume, of raw carrots and salad, but of three parts fresh horsemeat, two parts dog meal, and one part fresh milk.

This method of feeding had one disadvantage. The animals developed diarrhea. But as soon as the ferrets were returned to the standard laboratory diet, they again snapped and bit. Only a single 'pacific meal' sufficed, however, to make the savage animals gentle again. Whether similar procedures will work with other predators, such as wolves, lions, and tigers, is currently being investigated. For human beings, however, gentleness purchased at the price of continual diarrhea seems rather out of the question.

Domestic cocks[18] show a similar change to a peaceful disposition when they receive a special beverage—namely, alcohol. A contentious rooster will display distinctly maternal behavior toward chicks under the influence of a good shot of rye whiskey, and will even shelter the chicks at night. Presumably the alcohol suppresses aggressive behavior. The same effect has been noted in man. But often alcohol works in just the opposite way, sweeping away the inhibitions upon aggression.

One solution which seems acceptable in principle has been devised by the dwarfish Bushmen of the Kalahari Desert in South Africa. These nomad hunters necessarily have a highly developed aggressive instinct;

but in quarrels between man and man or tribe and tribe they curb that instinct by an ethical system which, while it contrasts sharply with ours, completely accords with the harsh conditions of their world. They have certain taboos which they inculcate into their children and which constitute a veritable system for training out quarrelsomeness.

To provoke a quarrel is the greatest crime there is in the world of the Kalahari Desert, says Danish anthropologist Jens Bjerre.[19] A tribe divided by bickering would not be able to survive under the almost inconceivably difficult conditions of the desert. Wars among the tribes would long ago have exterminated the race of Bushmen. Understanding this, these Stone Age savages behave far more rationally than civilized peoples.

Fights and even angry words are absolutely taboo. A Bushman who has been guilty of these is warned by the elders of the tribe; if he repeats his fault, he is expelled from the community. Such expulsion amounts practically to a death sentence, for a single human being can scarcely survive in the arid wilderness.

Even small children are punished severely for quarreling. They are made to accompany a hunting expedition of several days, which is anything but a pleasure. The result is that hardly anywhere in the world are there people who behave so peaceably toward one another as the Bushmen, in spite of their aggressive disposition. So much for the myth of the brutality of primitive peoples and of our nomadic forefathers.

The fact that Bushmen have been so successful in training quarrelsomeness out of small children reminds us of the experiments with mice and cats described at the beginning of this chapter. Their success shows the enormous effectiveness of personality-shaping measures when taken early enough.

The disposition to contentiousness is on the one hand inborn, fixed in the structure of the instincts, and subject to biological laws. But on the other hand the instinct is by no means rigid. Within a considerable compass its manifestations can be modified by environmental influences. Like anxiety, it can be either intensified or subdued. Those who would truly and deeply understand themselves and other human beings must be aware of both these fundamental elements.

7. Primate Children Learn Good Behavior

Status in the Jungle

Mothers are responsible for any lack of upbringing in their children. This commonly acknowledged principle applies not only to human beings but also to small South American squirrel monkeys known as death's-heads or saimiris.

The miscreant was young Wastl, who was only four weeks old. In size barely twice a man's thumb, the baby monkey rode on his mother's back, clinging tightly to her fur. From sheer ignorance of monkey manners, he reached out and touched the boss of the band. For these monkeys, such an act is as unseemly as it would be for a German to tug at the Chancellor's jacket to get his attention. But the surprising outcome was that the monkey boss whose dignity was offended did not turn upon the baby, but upon the mother.

Saimiris live in groups of up to several hundred animals in the jungles along the courses of the Amazon and Orinoco rivers in South America. Their social life is strictly regulated. Every infant must gradually learn what the community regards as good behavior. Professor Detlev Ploog,[1, 2] and his associate Sigrid Hopf studied the monkey methods of child rearing at the Max Planck Institute for Psychiatry in Munich.

In the first four weeks of its life the baby is allowed to do anything it pleases. It clings constantly to its mother and is to all intents and purposes a part of her. From its fifth week of life on, the mother begins to take measures to educate it. This involves a certain degree of sternness, for she must teach the baby to look around the world by itself occasionally and not always be hanging to her 'apron strings'. For a full year she must gradually wean the child, and with each passing week the education grows stricter. There is one special routine which is part of this training. The mother brushes the baby off her and pushes it away. Shortly afterward she coaxes it to come to her by calling. But as soon as it comes hopping toward her, she evades it. This is repeated several times, until at last she permits it to suckle. The baby is thus encouraged to act independently and deal with other members of the community. The baby itself is beginning to be eager to 'discover the world', and the mother speeds the process.

When a saimiri baby is eight weeks old, it begins to be treated as an individual. Only then do other members of the group make it responsible for its pranks. Impertinences toward the boss are no longer

A seven weeks old saimiri adopts this posture intended to impress an adult. (After H. Kacher.)

overlooked. The boss threatens the baby with increasing frequency and vigor. He shows his teeth, snarls, and even charges roughly at the 'fresh kid'. The severest tutorial measure is a bloodless nip. It hardly hurts, but the moral effect is tremendous: the baby screeches loudly and tries rather piteously to put on a bold front.

When Wastl was three months old he was sent into the 'kindergarten' where the other monkey children were romping—each one under the supervision of a 'governess', a female who had no young of her own. As often happens when a child is introduced into a new social group, in his very first day Wastl fell to fighting with an older ruffian and got the worst of it—naturally when his governess was not around to supervise. But Wastl was not going to take defeat tamely. He complained about his enemy.

Interestingly enough, saimiris handle their grievances in a far more useful fashion than human children. In the same situation a human child runs to his own mother. She in turn rushes to the miscreant's mother, who of course defends her darling—and a nasty quarrel among neighbors may go on for weeks, long after the children themselves have patched the whole thing up. But saimiris go about the matter differently. They themselves go to complain to their rough playmate's mother. That usually suffices to restore peace at once.

An actual exchange of blows rarely occurs among the saimiris. These monkeys usually settle their dispute merely by making threatening gestures and putting on airs. In general, those who put on airs get more out of life—especially in monkey society. It is ludicrous to see these Tom Thumbs trying to intimidate adults by 'posturing' when they are barely two days old. They 'spread themselves' in the most literal sense of the word: spreading one bent leg to the side and displaying their genitals. Even their big toes are spread apart from their other toes.

This posturing is used in a remarkable number of situations. For one thing, the gesture regulates life by averting serious fighting. The phylogenetic origin of the pose is obviously to be found in sexual behavior. But it has become ritualized as an independent instinctual act, and has acquired a nonsexual coloration.

Depending on who impresses whom, and when and in what

situation, the pose can have the following meanings: showing superiority to the other, cowing him in order to avoid a fight, defiance, standing up to or defending the self against a superior, pressing a demand, protesting, or complaining. Sometimes two rivals carry on lengthy 'posturing duels' with each other until one or the other of them is 'morally' defeated.

The male also postures in his courtship of the female. But by similar posturing the female can reject a suitor. Adults use the pose to reprove children and children to demand something of adults. Younger and weaker monkeys use it to gain status or to express disappointment. Finally the pose also serves as a greeting to a new arrival, to the human attendant, and even to a monkey's own mirror image.

In this case the mirror image is regarded as another monkey who has to be put down; this may be an indication that the animal imagines itself to be bigger than his actual stature as he sees it in the mirror. If so, vanity also exists in the animal kingdom.

It is hardly surprising, in the light of all this, that the babies will play at showing off the way puppies tussle. They practice the gesture most frequently toward adults, especially toward the boss and their mothers. If the baby is still young, the superior lets the 'effrontery' pass; he or she merely sniffs at the little fellow, the way a human father, challenged by his son, might soothingly place his hand on the boy's shoulder.

Adolescent monkeys, however, are reproved for such impudence. The adult grabs the youngster from behind, at the hip. He thus makes clear that posturing alone does not convey superiority. If the juvenile delinquent persists in his impertinence, the adult saimiri also reaches for the youth's head and vigorously forces it down, or pulls the rebel's ears.

Once a monkey has attained puberty, this kind of treatment involves serious problems. If he is 'ordered around' too much by an adult, he has a craving to posture, out of sheer defiance. At the same time he does not quite dare. But when he can no longer contain himself, he turns his back on the tyrannical elder and postures into space—just as a human 'bad boy' after a paternal reproof will stick out his tongue after he is safely out of sight.

There are play situations in which children are permitted to posture to adults and even to the boss. But on these occasions monkeys large and small must make a sign of friendliness to indicate that the posturing is not to be taken seriously. Dr. Peter Winter,[3] an associate of Professor Ploog, has discovered that this is done through a sound, that he calls 'play squeaking'. These little monkeys do not convey emotional tone by facial expression. Therefore a sound is used to neutralize the aggressive character of the posturing behavior.

The monkey grows up in two or three years, and learns what he may and may not do, what his position in the community is, and how he must behave toward others in order to live in peace. To learn all that requires a good measure of social intelligence.

To test the importance of the process of maturing, Dr. Ploog introduced a newcomer into an enclosure with a socially established group. It became evident that the first few minutes are crucial and decide whether the newcomer will be accepted peacefully into the community or whether there will be strife. Moreover, all this seems to hinge on the nature of the upbringing the monkey received in childhood.

The personal relationship of the newcomer to each individual member of the group depends almost entirely on the first impression he makes. As is true for men, there is liking, hostility, and indifference at first sight among the monkeys. Among saimiris, however, hostility does not necessarily lead to fighting, not even in the form of a posturing duel. Rather it is expressed by an avoidance of social contact. Here is an elementary sane procedure for keeping the peace.

But these matters take a different turn in a monkey group which has no such social mechanisms for regulating the community life. Witness an experiment arranged by Professor Ploog and Dr. Rolf Costell, who present the following account:

The group of Blackheads (so called because we marked their heads with a black stain) consisted of five males and six females; the group of Yellowheads of only three males, but thirteen females. When we opened the connecting door between the two cages, a loud and long-lasting battle of screaming immediately began. Both bosses postured continually toward one another. When this resulted in no decision, a fight began in which the females intervened. Finally the superior force of Yellowheads drove the Blackhead group out of their sunnier territory. . . .

The subsequent merging process took several weeks, during which the continuous posturing of all participants revealed the extraordinary dynamism of social group-formation. There were also some highly interesting incidental aspects. For example, after it had become clear that the Yellow boss was the strongest member of the merging group as a whole, he entered into an alliance with the highest-ranking monkey of the Blackheads, who now had to accept second position. Both so thoroughly repressed the former second-in-command of the Yellowheads that he dropped to the lowest rank in the entire group and became the 'whipping boy'.

As it turned out, it was all of five weeks before conditions stabilized. By this time each individual had gradually adjusted to the others and found his position in the new community. The adjustments involved considerable changes in the personalities of individuals, as Detlev Ploog could demonstrate on the basis of the comprehensive behavioral profiles drawn up before and after the merger. But the tensions abated and gave way to the ordinarily peaceful community life of these monkeys.

The personality of a monkey might change from loner to sociable pal if he happened to strike up a friendship with one of the new monkeys belonging to the other party, and thereby gained admission to their circle. On the other hand, a cheeky and obstreperous little fellow quieted down considerably when he realized that he could not make as much of an impression on the strangers as he had been accustomed to doing in his own group.

In every case, the status any one saimiri was granted by the others depended very heavily on his skill at posturing. But clever and political use of this technique is in its turn dependent to a high degree on childhood upbringing.

The facts are inescapable: among monkeys, too, there are well and badly brought-up children; the results depend on the mothers' pedagogical abilities. What is botched in childhood can never be redeemed in a monkey's lifetime. This principle is in fact so sweeping that a distinct class-consciousness is bred in the juvenile saimiris. The offspring of parents low in rank can never rise to a high position in monkey society—not even by virtue of outstanding physical strength. Such facts as these, hitherto totally unknown in the life of animals, were discovered by Japanese zoologists[4, 5] in troops of red-faced macaques.

These animals, which are about the size of rhesus monkeys, live in troops of one hundred to three hundred individuals on the islands of Japan. Each troop claims a territory of from one to five square kilometers, the boundaries being coasts, streams, mountain chains, and human towns or villages. Within their camping grounds the spot where each individual eats, plays, and rests is subject to an etiquette as strict as that governing diplomatic functions. The females and children have their places in the center. Round about them the males form several concentric rings.

The lower a monkey is in social status, the farther toward the outside he must stay. The adolescents are banished to the outermost and most dangerous ring, where enemies are near and females far.

Only a few males very high in rank enjoy the privilege of entering the center where the females sit. Among these are the boss and those males who occupy the innermost ring, directly adjacent to the females' area. Thus rank is of crucial importance for the destiny of every individual monkey; it governs all he is allowed to do and all the pleasures that are permitted him or withheld from him. In many animal societies, among chickens, wolves, or chamois, for example, rank is determined by battle. But among Japanese monkeys a totally different system prevails. Here there is no fighting. Instead the social status of the parents is transferred as a matter of course to the children—like titles of nobility among human beings.

The principles of inherited status follow rigid laws. At marriage the female automatically assumes the rank of her mate. The preferred wife of the troop leader, for example, is superior to the second in command. We have seen similar phenomena among ravens, daws, geese, and several other animals. But the system of privilege goes one step farther among the red-faced macaques since the offspring also assume the rank of their parents. Among brothers and sisters, moreover, the youngest child at any given time assumes the highest rank.

In this way the children of the 'upper classes' grow accustomed from infancy to domineering behavior. If, for example, two monkeys

Pride of class in the jungle is displayed by the Japanese red-faced macaque. This animal has a strikingly red facial coloration and lives in large troops.

are interested in the same food, they do not fight over it. Instead the higher-ranking one promptly postures, thus threatening his rival. If the lower-ranking monkey wishes to avoid a painful beating, he must try to appease the other. This he can do only by adopting the female posture of copulation. The human observer finds it highly amusing when in the course of such a dispute a lower-ranking but muscular male gets down on his knees before a higher-ranking but small female. Naturally, the children owe their authority solely to the alertness of their parents, who take care that the inherited rank is respected.

The 'personality' that grows up in this way necessarily develops haughtiness that we may punningly call first-class. But after all, the reputation of human beings also is founded sometimes more on their demeanor than on their actual accomplishments. It is therefore not surprising that a young monkey of the aristocratic class assumes superiority to his fellows even when his parents are no longer guarding his 'dignity'. Solely by showing off, he climbs to the rank of a kind of subleader in the 'club' of adolescents. On the other hand, the monkeys that have been taught the gesture of submission by their low-ranking parents rarely achieve any status.

One of the Japanese scientists, Dr. Yamada, observed how after the death of the troop leader an aristocratic youth, who had previously been the leader of a band of young monkeys his own age, was accepted as successor and commander of more than three hundred monkeys. There was no fighting. Unfortunately, there is no evidence that the new leader was a son of the deceased 'king'. Had that been the case, we would actually have something akin to a hereditary monarchy among these animals.

The fate of 'young ladies of good family' is something else again. It is the custom of certain macaque troops in Japan for the females to leave their mothers as soon as they are grown. Their future rank is then determined by the male with whom they associate. In these troops the females never have much influence over their mates. The males treat them quite roughly, beating and biting them for trivial causes.

In other troops, however, the females have fought for and won equal rights. These troops present a remarkable example of the formation and persistence of traditions. There are such troops on the small island of Koshima in southern Japan, and on Minootani Mountain near Osaka in Central Japan. Among these monkeys, the greatest good fortune of a grown female is to raise as many daughters as possible. For by custom the daughters remain with their mother even after they are grown. She is the originator of a matriarchate that seems unique in the animal kingdom, and she retains her power even when she has become a grandmother and great-grandmother. In this matriarchate husbands are merely tolerated. Because the daughters remain with the mothers, a powerful clique of women arises, so strong that it can quash any ugliness on the part of the husbands and sons-in-law.

Can Electroshock Replace Mother Love?

'Psychologically speaking, the child is the father of the man,' Sigmund Freud has observed. 'The impressions that it receives in the first months and years of its existence are of decisive importance for its whole life.' During his own lifetime, however, Freud was unable to provide experimental proof for this fundamental concept of his theory. Only since the end of the nineteen fifties have psychologists and behavioral researchers been able to supply biological facts that suggest the need for checking and revising earlier ideas.

The view that childhood impressions leave an indelible imprint upon a living organism, and determine its later destinies, hardly encounters any opposition nowadays, if formulated in this highly general manner. But who is prepared to draw the necessary conclusions? The specifics, of course, are another matter and may well encounter emotional resistance—but the facts cannot be blinked.

To this day we do not know precisely what mother love is. The child's love for the mother has often in the past been regarded as a mystical force. More realistically, it has been defined as the result of 'training with food rewards'. Both these views have proved to be false.

Let us for a moment consider one of the first actions of a newborn human child: crying. As soon as the mother holds the wailing infant to her breast, it quiets down—even without suckling, unless it is terribly hungry. But it is not the total presence of the mother that has the quieting effect. A single factor does the trick: the sound of her heart beating quietly!

At the suggestion of Professor Jonas Salk,[6] who developed the vaccine for poliomyelitis, this discovery was put to the test in a New York maternity hospital:

> The gentle dub-dupp, dub-dupp of a mother's quiet heartbeat was played from a tape on a loudspeaker system in a room full of new-

born babies. At the picture window outside, a hospital attendant kept a checklist on the babies in their baskets. Most of them fell asleep soon after the heartbeats began to be heard; the rest seemed content with themselves and the world. When the recording of the heart sounds stopped, many of the babies awoke within a few minutes, and several of them began to cry.

Then another tape was played. It reproduced the rapid heartbeat of an agitated woman. The sound was no louder, but all the sleeping infants awoke at once. All the children grew tense, as if they were afraid. When the first tape was replayed, peace spread through the nursery once more.

The 'mood music' of the maternal heartbeat affects the child's organism as well as its emotions: the baby's heart activity also changes. The faster the mother's heart beats, the more excitedly the newborn infant's pulses trip. This compulsion to be affected by an imposed rhythm continues, in a broader sense, throughout the whole of later life. Rhythm is the fundamental element of music. Military bands play at a tempo that is somewhat faster than the normal pace of the heart. Consequently they 'make the heart beat higher'.

The influence of the maternal heartbeat upon the baby may also account for a remarkable physiologic fact. African infants who are carried day in and day out in a sling on their mothers' backs, where they can hear the heartbeat, are far ahead of babies of civilized peoples in all aspects of development. This was ascertained by Drs. R. Dean and Marcelle Geber[7] in studies of the Kampala tribe in Uganda.

But from the sixth month on the whole picture changes. Then the baby is weaned, and following the ancient custom of the tribe is given into the care of one of its grandmothers. From that point on the children vegetate—there is no other word for it—in a monotonous environment. They are subject to harsh control, given no love, stimulus or playthings, fed on little more than potatoes, and live in sad and dirty surroundings. Their mental development seems suddenly to be paralyzed. Most three-year-olds appear utterly apathetic and indifferent.

In terms of heredity, blacks are no more and no less intelligent than whites. But given such methods of child rearing, we cannot expect any high mental achievements of these people when they reach adulthood.

What kind of rearing creates the strongest attachment on the part of children to those who take care of them—strictness, love, or capriciousness? Hoping to obtain some clues from the behavior of dogs, Dr. A. E. Fisher[8] divided a group of whelps the same age into three groups and 'reared' each group by a different method.

He invariably rewarded the pups of Group A whenever they approached him by caressing them and giving them food. The animals were allowed to lick his face and tear the legs of his trousers. Whatever they did, he responded affectionately. On the other hand, he always sharply rejected the puppies of Group B whenever they tried to touch him. He punished every slightest misdeed with blows. The animals received food only in his absence, and then it was pushed to them through a hole. The whelps in Group C received caresses and blows in a quite arbitrary manner. Sometimes he would allow them to lick him, then again he would push them away without any apparent reason.

The results were surprising. For it was not the dogs raised by the kindly method who in later life developed the greatest attachment to their keeper. Affection on the part of the dogs did not grow out of their having received constant affection during their puppyhood, and certainly not out of pedantic discipline and sternness. Rather, the dogs that remained most devoted to their master were the ones that had been sometimes beaten and sometimes caressed.

This result appears illogical only in the light of human principles. For animal parents are extremely capricious. Sometimes they want to be left alone and repulse their offspring; at other times they are in the mood for play and a bit of roughhouse. In this respect, too, nature has taken a highly meaningful course. Undoubtedly there are some elements of this response in human children also; but there is one fundamental difference. Human children have a fine feeling for justice. Therefore, in our child rearing, justice must take the place of capriciousness.

There is another important element. An animal baby can put up with harsh measures if it knows where it belongs. City-dwelling animal lovers commit a multitude of sins against this principle. One day they pamper their dogs, next day they leave the animal for several weeks in a boarding kennel. Then again they sell or give the poor creature away, and in its new home the same anguish begins all over again. That is much greater cruelty to animals than a barrage of slaps. No wonder that many dogs become neurotic—literally neurotic. That is no joke: there are specialized animal psychiatrists in a number of cities who are engaged in treating just such conditions.

What are the characteristics of a mother that particularly matter to a child? Professor Harry F. Harlow,[9, 10] has carried out some famous experiments with rhesus monkeys at the Primate Laboratory of the University of Wisconsin in an attempt to elucidate this question.

The psychologist came to the following conclusion: the most essential characteristic of a monkey mother is a soft something to which the baby can cling. It apparently does not matter whether this soft something is a living mother or a doll. A large doll with a wooden head, glass eyes and a soft cloth chest, from which the rubber nipples of two baby bottles protruded instead of teats, seemed at first sight to be a full equivalent for the natural mother as far as the baby was concerned.

If the psychologist introduced a wind-up teddy bear into the baby monkey's cage, one which could move with a rattling sound, and beat a drum, and if no mother doll were present, the monkey crept into the corner in terror, and for hours did not dare to move. But if the doll was inside the cage, the baby monkey took one frightened leap away from the unknown toy to the lifeless mother surrogate, clung tightly to it, rubbed his belly against its cloth, then climbed up somewhat higher to look into the 'mother's' glass eyes, and then, seeing no signs of fear on the mother's part, returned cautiously to make an exploration. Finally the baby monkey examined the teddy bear closely and began to play with it.

Thus the sham mother gave the baby a sense of security and banished its fear.

The monkeys which were separated from their real mothers a few hours after birth and given the company of dolls throve much better than those under the care of their natural mothers, according to

Harlow. They grew faster and were considerably healthier. Infant mortality was rarer among them. Can we draw any conclusions from these facts?

If we did, we would surely be mistaken. Three years later, when the rhesus monkeys that had been raised in this unnatural way reached

Can a cloth doll with a wooden head and glass eyes serve as a complete substitute for a rhesus baby's mother?

adulthood, the disastrous results of this kind of rearing became apparent. The adults proved to be psychically warped and completely asocial. They sat apathetically in a corner of their cages and stared into space, or else they prowled restlessly back and forth. Then they cradled their heads on their arms and rocked back and forth for hours. Or else they pinched themselves hundreds of times a day on the same part of their bodies, until they bled. When badgered by a human being, they bit their own bodies—in a grotesque perversion of the normal aggressive instinct. They were incapable of making any contact with others of their kind, incapable of playing with other monkeys or reacting to them in any way. The opposite sex left them cold.

While the substitute mother had served well enough for the monkey's physical development, it had been unable to supply what was needed for the psychic, nervous, and hormonal processes of maturation.

When the asocial females were artificially inseminated and bore young, they did not know what to do with them. The cries of the babies and their helpless little forms did not move the mothers. Indifferently, they let the babies lie and scream. Had the human experimenter not intervened, the little creatures would have died.

Being a mother is even harder than becoming a mother. In contrast to many other animals, monkeys must, like human beings, learn the ins and outs of maternal behavior. Jane van Lawick-Goodall (*see* Chapter 1, note 22) relates amusingly how four-year-old female chimpanzees 'play mama'. They are highly interested in everything their mother does with their brothers and sisters. Dr. Wolfgang Wickler[11] reports:

> The older daughter attempts to caress the baby, or tugs cautiously at its arm or leg. The mother pushes the still unskilled sister gently but firmly aside. When the infant is somewhat older, the bigger child is permitted to hold it now and then. But the mother is always nearby, paying close attention, and takes the baby back as soon as it cries. When the baby begins to crawl around, the elder sister watches it and jealously fends off other inquisitive age-mates. She imitates some of the games the mother plays with her infant and also tries to feed the baby from mouth to mouth with prechewed food, as the mother does. When the baby seizes hold of her hair and tugs, she discreetly tries to make it let go. If unable to do so, she carries it back to the mother.

Thus the young female chimpanzee grows into its future role as a mother by playing. Without such teaching by example within the family circle, she will not learn the necessary techniques. We have seen examples of this ignorance in the case of many zoo animals that have been kept by themselves. The mother of the Basel Zoo's gorilla baby Goma, which became a great favorite with the public, was completely unfamiliar with baby care. She had to be shown by human attendants how to handle her first child. Then she was able to raise her second by herself.

But aside from the details of mothering, must all social behavior be learned? Or could the sad story of the rhesus monkeys that grew up with dolls be traced to a nerve maturation process that did not take its proper course? If so, can this process be set in motion by artificial stimuli which will replace social contacts between animals? Such a prospect sounds like the fantasies of science fiction. Nevertheless Professor Seymour Levine[12] in 1960 conducted experiments at London University which seem to point to some such possibility.

Professor Levine is a psychiatrist. His subjects were baby rats. All of them were kept in individual cages without a mother, and were divided into two categories. The experimenter administered an electric shock once a day to the animals of the first group. He left the others undisturbed during the sensitive period of infancy. In the latter group, the results confirmed the outcome of Harlow's experiments with rhesus monkeys: when they reached adulthood the rats which had never been exposed to normal stimuli, and which had experienced nothing but their four walls, behaved as abnormally as the motherless monkeys. But the daily shock administered to the rats of the other group seemed to have had a beneficial effect upon their emotional development. Professor Levine informs us that in their behavior they could not be distinguished from rats that had grown naturally in the care of their mothers.

Can electroshock then replace mother love? Surely this is a fantastic notion. And such a disquieting conclusion is at least premature, since as yet only the activity-anxiety complex, not the whole scale of behavior of the rats, has been examined. But the experiment at least proves that psychic stimulus is an essential part of a baby's early experience.

Professor Levine found that stimulation during the first sixteen days of a rat's life triggers many series of hormones in the organism. It hastens and varies the maturation process in the central nervous system. One manifestation of this process is an increasing supply of cholesterol to the brain. The animals open their eyes sooner than the unshocked rats. They are able to coordinate their leg movements sooner, have a greater resistance to diseases, and are free of exaggerated timidity.

Experiences in infancy, then, affect the emotional life of grown animals by way of the hormones and the nervous system. Professor Victor H. Denenberg,[13] the American psychologist, has undertaken a similar series of experiments at Purdue University which show that activity, aggression and enterprise, or anxiety and timidity, in an adult rat are much more strongly determined by the environmental influences in infancy than by hereditary disposition.

All the animals in the experiments we have described had virtually never seen their mothers. They did not know that such a thing as a mother existed. But cases in which a child knows its mother and then loses her because of some tragic event are much closer to reality. If it can survive at all, what is the effect of this critical loss upon the further development of an orphaned young animal? In 1966 two American psychiatrists, Professors Charles Kaufman and Leonard A. Rosenblum,[14] examined this question.

For the first experiment the scientists kept a select company of pig-tailed macaques, large relatives of the rhesus monkey, in a spacious enclosure at the laboratory. There were four mothers, each with a baby, a childless female, and a male, the father of all the offspring. As soon as a baby reached the age of five months, a situation corresponding to the sudden death of the mother was created. The mother was removed from the enclosure, while the young monkey remained with the group.

Things at once took a dramatic turn. During the first twenty-four to thirty-six hours, all the motherless monkeys exhibited a high degree of excitement. They constantly sounded their plaintive cry of distress. They wandered about, obviously in search of their mothers, and broke into purposeless bouts of activity which just as suddenly gave way to rigid postures.

On the second day this behavior changed completely. All outwardly recognizable activity ceased. The little fellows crouched in a corner with heads lowered and shoulders bowed. But in spite of their severe depression the 'orphans' now and then made shy attempts at contact with other members of the group. But the latter met them only with hostility, making threatening gestures and pushing them away. Thus

The pig-tailed macaques which inhabit the jungles of Indonesia show no pity for orphans. Motherless babies are doomed to die.

the wretched situation of the abandoned offspring could be attributed not only to their own behavior; their 'society' also rejected them.

The monkeys of another group, with whom the same procedure was followed, reacted in entirely different fashion. These were the bonnet macaques. The species is closely related to the pig-tailed macaques, but these monkeys with their curious topknot at once befriended the abandoned young ones. Other mothers adopted and suckled them, other babies continued to play with them, and so the orphans gradually recovered from their initial symptoms of severe depression.

Was the affection of the others a total substitute for mother love for these monkey children? The American scientists found a surprising answer to this question when they reunited mother and child after a separation of four weeks. Among the pig-tailed monkeys of the first experiment, the course of events was exactly what had been expected. There was a scene of heartrending greeting, and thereafter mother and child were all but inseparable.

It might be expected that the bonnet macaque orphans of the second series, who had apparently been thoroughly consoled for their loss by other members of the troop, would receive their restored mothers fairly indifferently after four weeks of separation. But that was not how it turned out. The rejoicing on the part of the baby bonnet macaques was just as vehement as it had been among the rejected, heartbroken pig-tailed macaques of the first experiment. Hence mother love must mean more to these animals than merely food, someone to play with, and safety.

The seeming indifference of young bonnet macaques to the loss of their mothers is reminiscent of the 'apathy' that human children display in similar situations. The common opinion that young children scarcely feel any grief over the death of mother or father, or that they cannot really grasp what has happened, has been called totally absurd by one team of eight American psychoanalysts. In reality the loss affects the child far more strongly than a mourning adult. Adults simply do not understand the supposedly emotionally cold child.

In adults, too, grief is a combination of anxiety, helplessness, disappointment, and psychic shock, declares Dr. Joan Simmons,[15] head of the Chicago team. Mourning manifests elements of mental rigidity, she points out. It is an act of unconscious self-defense, when experiences become so overwhelming that the psyche cannot endure them all at once.

Such terrible experiences produce a much more lasting effect in a child. The mental rigidity becomes a kind of numbness and can develop into a lifelong trauma unless some experienced person gives the child help. The team of psychoanalysts based their opinions on fifty adult patients who had all suffered severe and lasting disturbances in their emotional life due to the early loss of parents.

The child in his relationships with others does not deny the fact of

Bonnet macaques of the island of Celebes affectionately take charge of orphaned young. Nevertheless they cannot prevent the orphans from suffering psychological injury.

the loss, but represses almost all the emotion connected with it. Much like a mourning adult, a child escapes into a world of memories where it can pretend that the mother or father is still alive. To preserve this fiction the child expends a great deal of mental energy which is then not available when it is needed for other aspects of the process of psychic development.

Cambridge Professor R. A. Hinde[16] has come to the same conclusion after experiments with rhesus monkeys. Monkeys which had been separated from their mothers for only six days in their infancy revealed severe psychic defects in later life.

In addition, a further surprising phenomenon emerged from Kaufman's and Rosenblum's experiments with pig-tailed macaques and bonnet macaques. The scientists had deliberately fixed the time of reunion between mother and child for the seventh month of the young monkey's life. At this period monkey mothers normally wean their offspring, teach them to fend for themselves, and finally repulse them.

But these developments did not take place with the animals in either of the two series of experiments. After the painful separation, there were no longer any limits to the affection between mother and child. Henceforth they stayed close together all the time. The separation had tightened the bond.

Findings such as these are highly pertinent to the mental and psychological development of human children. They are certainly important to boarding schools and orphanages. At the very least, there must be some amendment to the rules of children's hospitals which restrict parental visits to twice a week—even though more frequent visiting involves the danger of many painful leavetakings. Above all, parents ought to consider carefully what they are doing when they place a child under ten years of age in an infants' home or a boarding school.

It has by now become apparent that there are a great many tangible and intangible biological and psychological factors in the normal course of a child's development. Man is not a constant creature with an emotional life fixed once and for all, a personality established for all time, and mental abilities and intelligence programmed into him from birth on. Whether a human being will be intelligent or stupid later in life does not depend entirely on inheritance. To a very high degree, the psychic influences and mental stimuli that affect him from earliest childhood will govern the quality of his intelligence.

8. The Three Pillars of Intelligence

Animals Invent Tools

What distinguishes man from animal? Only a few years ago anthropologists might have answered: Man is the only being intelligent enough to invent, produce, and employ tools.

In the meantime, however, zoologists have called this idea into question. They are discovering more and more examples of the use of tools by animals. Certainly, in many cases the animals are probably acting instinctively, and in other cases they simply find the tools, such as stones, but do not invent them. In a number of primate societies, however, there are undoubtedly occasional individuals of 'genius' who become true inventors, helping their troops to achieve 'technological progress'. American behavioral scientists consequently speak of man nowadays only as the 'paramount toolmaker.' Thus the difference between man and animal becomes only one of degree—at least in this particular realm.

Let us quickly run through the abilities of animals in this area. We discover the first users of tools in the class of insects, among several American species of the sand wasp *Ammophila*.[1] After the female wasp has dragged a paralyzed caterpillar into the underground chamber she herself has dug, and has laid an egg on it, she closes the entrance with earth to protect it from other parasites. To keep wind and rain from exposing the chamber again, the small insect pounds the earth as hard as she can, as though she were paving a road. She uses a tool for this purpose: a pebble held between her mandibles.

The weaver ant, *Oecophylla smaragdina*,[2] uses its own silk-producing larvae as weaving shuttles for nest building. While a number of the ants pull leaves into the proper position, other workers take the larvae which are ready to spin their cocoons and squeeze them until a liquid silk secretion is forced out of them like the contents of a tube of liquid glue. With this secretion they stick the leaves together.

Other ant colonies use full-fledged members of the society as tools. In the storerooms of the famous honey ants[3] as many as six hundred of the insects cling to the ceiling as living honeypots and allow themselves to be stuffed with honey until their abdomens have swollen to the size of a pea. Colocopsis ants have even developed a special caste of 'living doors'. The huge heads of the workers resemble bottle corks. They squeeze their heads into the entrance, sealing it hermetically. Legitimate callers must tap their identification code upon it before the 'door' is opened for them.

Through no choice of their own, ants occasionally serve various birds as tools. When, for example, chaffinches[4] feel an itch somewhere on their bodies and happen to see an ant, they pick up the insect in the tip of their beaks, place it under their feathers and let it pinch them at the right spot. This 'emmeting' as the scientists call it, has also been observed among song thrushes and missel thrushes, blackbirds, water-ouzels, and starlings,[5] and according to very recent researches is wide-spread in the avian world.

Many birds behave as if they were afraid of the ants and after the 'treatment' at once hurl them away. Other birds do not venture to touch ants at all and apparently are unfamiliar with this measure. Presumably, then, we are dealing here with learned behavior.

Some birds may even fashion tools for themselves. In New Guinea and neighbouring regions, for example, there are various bowerbirds[6] which for purposes of courtship build highly artistic arbors woven of thousands of twigs and grass-blades. Moreover they paint these structures in colors. This cannot be done without painter's implements, and the birds actually make themselves paintbrushes. The satin bowerbird

(*Ptilonorhynchus violaceus*) of Queensland shreds bits of bark for the purpose; the gardener bowerbird bundles together dried bits of leaves. For paint these feathered artists use blue-gray or pea-green saliva.

The woodpecker finch[7] and the mangrove finch[8] of the Galapagos Islands in the Pacific manipulate cactus thorns or small twigs so that they can use them as a woodpecker does its beak, to pick insects out of rotting wood.

In 1966 Jane and Hugo van Lawick-Goodall[9] reported an especially interesting case of the use of tools among birds.

The couple were working in the Serengeti Plain in Tanzania. The day before, a grass fire had swept over the plain, and among other damage had driven the ostriches from their nests. Now several vultures were standing around the abandoned eggs. Ostrich eggs weigh two to three pounds and have so hard a shell that a man can open them only with a hammer. The griffons and eared vultures, big as eagles, hacked away at the eggs with their powerful beaks, but in vain. They could not manage to crack a single egg.

Then two Egyptian vultures came floating down. These birds, no bigger than hens, are dwarfs in comparison to the other vultures. Nevertheless they solved the problem. At a distance of some ten yards from the eggs they picked out some smallish stones. With these in their beaks, they stalked up to the well-sealed dainties. Then they raised their beaks high and hurled the stones with all their might at the eggs. After repeated blows of the stones the eggs cracked and the meal could begin.

It is significant that the larger vultures watched the technique of their relatives, but were incapable of imitating. They once more hacked at the eggs with redoubled violence, but without any more success. None of them profited by the example. They simply lacked the necessary tool-using intelligence.

Australian buzzards[10] practice bombing. They carry stones up to the size of a hen's egg to a height of twelve or fifteen feet and let them drop upon an unguarded clutch of emu eggs.

Another 'Stone Age animal' is the jolly, playful sea otter of the Pacific Coast of North America. This animal uses stones as kitchen utensils, in order to enjoy special delicacies: oysters, mussels, crabs, sea snails, and sea urchins. In 1964 Dr. K. R. L. Hall and Dr. George B. Schaller[11] published one of the first scientific studies of their behavior.

The otter chooses a smooth stone about the size of a man's fist, tucks it under one armpit, and dives for an oyster. As soon as he comes to the surface with his armored prey, he turns belly-upward in the water, skillfully balances the stone on his chest, and holding the oyster in both hands pounds it against the stone anvil until it breaks open and can be eaten.

Schaller maintains that this may well be called an intelligent invention on the part of sea otters. Young otters at play scare one another by pounding stones together to make a loud noise. The use of stones as tools might have developed out of this playful impulse.

Such tricks, however, seem minor compared to the amazing achievements of beavers in hydraulic engineering.[12, 13] Witness the feats of a single pair of beavers who in a fifteen-month period felled 270 trees, including giants five feet in diameter and more than one hundred feet high. In addition the pair built three dams ranging in length from 130 feet to 200 feet, and set aside a supply of winter food.

There are, of course, limits to the intelligence of this superb engineer. Leonard Lee Rue III[14] comments:

Many people credit the beaver with more sagacity than this remarkable creature actually has. They claim it can fell a tree in any desired direction. This of course is not true, as can readily be determined by anyone who will take the trouble to investigate it. According to my own observations, I find that one out of every five trees that a beaver cuts cannot be used by the beaver because it becomes lodged against other standing trees and never falls to the ground, or else it falls over the tops of other trees already down and cannot be reached by the beaver. Most of the trees do fall in the water where beavers want them to, but that is because of good biological reasons and gravity, and not as the direct result of expert woodsmanship. Trees growing along streams will send out many branches on the side toward the stream where there is more sunlight.

When such trees are cut, they are heavier on the stream side and fall in that direction, a fact for which the beaver gets all the credit.

The beaver does not work like a woodsman with an axe, but chews his way around the trunk until there is only a thin sliver left in the middle. As soon as the tree begins to creak, all the beavers in the vicinity flee to safety. But accidents happen, and careless beavers are sometimes killed by a falling tree like careless human woodchoppers.

After the tree is down, the animals strip the bark from the trunk, remove the branches, and cut up the heavy wood into pieces of varying lengths. The longer the land route to the water is, the smaller the pieces. If the distance is too far, the beavers first make canals sometimes more than three hundred feet in length, along which they float the wood to the dam site. Sometimes stretches of these canals are tunneled through stony hills. Occasionally these water tunnels also serve as supplementary escape routes. It follows that beavers must have a maplike notion of their territory in their heads.

The dams fully deserve comparison with carefully planned engineering work. Dams seven hundred feet in length are not a rarity in Canada. One dam in Montana was 2,140 feet long. The record was a dam near Berlin, New Hampshire, no less than four thousand feet in length! If the current of the stream is weak, the beavers build their dams ruler-straight. But where the water pressure is strong they use a bowed baseline, as if they understood the laws of hydrostatics. In exceptionally swift streams they build one or more cofferdams above the main dam to tame the rushing water. If necessary, and if the stream is not too wide, they may turn it into an entirely new bed.

The beavers do not command the technique of ramming posts into the ground. Instead, they weigh down brush with heavy stones, and add more and more sticks to the structure. Interstices are filled with a 'mortar' made of mud and leaves.

In the direction against the stream the beavers build their embankment at an angle of 45°, while on the downstream side the slope is considerably steeper. The danger of high water is also taken into consideration: spillways are built into the dam which are only provisionally closed with light material and can be quickly opened.

To build their lodges, the beavers usually choose some bushes still protruding from the water in their pond. These serve to anchor the foundation, which consists of a circle of firmly compressed mud fifteen or more feet in diameter. It forms an artificial island rising about a foot above the surface of the water. Upon this a solid dome of branches is erected. Inside there are usually several rooms. Even two-storey houses have been found.

Every beaver citadel contains two hidden entrances about three feet below the surface of the water. These do not freeze over in winter and permit the beavers to enter and to bring in the wood which forms their food supply. The 'service entrance' has a special purpose: in case of danger an escape route free of obstructing pieces of wood must be available.

Beavers are perpetually engaged in a struggle against the swift silting up of their artificial ponds. The dams must be constantly raised or re-built. The result is that in the course of millennia millions upon millions of beavers have changed the face of the earth in large parts of North America. Rough valleys were transformed into fruitful meadowland. Then the white man came and decided that the beaver were harmful animals, aside from their valuable skins. He slaughtered them by the hundreds of thousands annually. The dams decayed, and fruitful land became stony waste once again.

And something very strange happened to the beaver itself. Where only a few beavers were left, they lost their engineering skills. For the survivors, there was room enough to live on natural lakes. Population pressure no longer compelled them to create their artificial ponds. Even when they were placed under protection and began increasing once more, they were initially incapable of complicated architectural works. But slowly, in the course of recent decades, they have regained their old abilities.

This fact, and the conclusion of the Swiss anatomist Professor Pilleri that the beaver has 'an unusually strongly developed cerebral cortex', suggests that the astonishing achievements of beavers are not a matter of instinctive ability, but represent learned behavior acquired by laborious experimentation and passed on to the offspring and possibly to neighbors as well—just as is the case among human beings. We may begin to sympathize with the belief of the North American Indians in whose cosmogony man was descended not from the ape, but from the beaver.

For it is an embarrassing fact that apes and monkeys are quite without talent for architecture. Moreover, they have not come nearly so far in the use of 'Stone Age' tools as the sea otter or the Egyptian vulture. The widespread notion that apes use stones or coconuts as missiles must be received with skepticism. Two American primatologists, Dr. Washburn and Dr. DeVore (*see* Chapter 1, note 18), have observed that when baboons fly into a rage they will hurl everything they can lay hands on in all directions. But they almost never hit the object of their anger. In fact, no intention to hit anything can be demonstrated. The whole spectacle is merely a display of strength, a tantrum meant to intimidate some antagonist.

Chimpanzees, however, ultimately developed the use of weapons out of this kind of posturing behavior, as we have seen at the beginning of this book. Dr. Adriaan Kortland (*see* Chapter 1, note 8) also reports that man's closest relative frequently threatens an enemy by shaking and even uprooting saplings. The animal probably meant to swing the tree or branches impressively in the general direction of the enemy, making a loud, swishing noise. But it often happens that the tree breaks and the ape unexpectedly finds itself with a large club in its hands. From this point it is only a logical step to running at the enemy with the sapling which has now become a weapon, and to hit him with it.

A less spectacular invention, but one that implies a further development in intelligence, is the use of a feeding utensil among savanna chimpanzees. Jane van Lawick-Goodall (*see* Chapter 1, note 22) observed the animals deliberately converting small twigs into a kind of fishing rod. First they stripped the leaves from a twig and shortened it to a length of about a foot. Then they went in search of a termite hill. They scratched open the holes of the insect fortress, thrust the sticks in, and waited until the termite soldiers had bitten hard into the wood and were clinging to it like bulldogs. Then they effortlessly pulled the sticks out and licked off the termites with obvious pleasure. Some chimpanzees spent as much as two hours at this activity.

Once upon a time some 'genius' of a chimpanzee probably devised this technique. Thereafter it made the rounds of the chimpanzee troop by imitation. Chimpanzees also thrust larger sticks into beehives in order to lick off the 'honey on a stick'. During the dry season they mash leaves into a kind of sponge with which they dip drinking water out of narrow cracks. After juicy meals they will also wipe their hands on 'napkins' of large leaves—or even use such leaves as toilet paper.

Only very rarely do men have the opportunity to witness such ape inventions. So it was a lucky day at the Munich Zoo when a chimpanzee (see Chapter 2, note 39), who had often watched the attendant unlock the cage door with a key, actually imitated the human action. Moreover, he made himself a key by chewing the end of a stick until it fitted the simple lock, and opened his own cage!

Unfortunately a safety lock was quickly installed, thus defeating the ambitions of the clever animal. Otherwise the zoo people might have seen whether and how the animal locksmith communicated his achievement to other chimpanzees.

For primates can and do pass on such learned accomplishments. It was reserved for Japanese scientists to watch every detail of this interesting process over a period of ten years. The primates were the natives of their islands, the red-faced macaques.

One day in the autumn of 1953 a monkey female aged one and a half years, to whom the Japanese observers had given the name of Imo, made an incontrovertible discovery. Along a lakeshore she had found a sweet potato covered with sand and had dipped it in the water— probably by pure chance at first. The sand was washed away. The monkey helped the process along somewhat by rubbing it with her hands. Thus she washed the sweet potato clean and evidently noticed that it tasted better washed than dirty.

By that act Imo, as the zoologist Dr. Masao Kawai tells the story,[15] founded a higher monkey culture which was later to make the small island of Koshima famous.

A month later one of Imo's playmates also began washing sweet potatoes before eating them. Four months later Imo's mother had actually learned this bit of 'household' lore from her daughter. Gradually this behavior spread through the troop among the mothers and babies, age-mates, and playmates. By 1957, four years after Imo's first washing of a potato, fifteen macaques had adopted the trick. Only the children and the females learned it, however. The males were evidently too proud or too obstinate to learn anything from their juniors. It is rumored that human beings behave somewhat similarly.

During the early period, therefore, the new achievement was passed only from the children to their mothers and from younger monkeys, Imo's playmates, to slightly older ones. Later on, however, when the practice of washing sweet potatoes was already fairly widespread, the mothers taught it to the children that had been born to them in the meantime. In this way, after a total of ten years, forty-two of the fifty-nine monkeys in the troop made a habit of washing unclean food, the invention of a young female. Only the old males still resisted the newfangled custom—and remained steadfast until they died.

Later still, the Japanese macaques refined the method somewhat. They no longer washed the sweet potatoes in the fresh water of a lake or brook, but in the salt water at the seashore. Perhaps they preferred their potatoes salted.

These same monkeys learned a special technique for enjoying grain. Grains of wheat strewn on the ground became mixed with soil or sand. At first the animals had picked the grains out of the sand one by one. Later they simply tossed a handful into the water. The sand sank to the bottom while the light grain floated. The monkeys had only to scoop the grain from the surface of the water and eat it with perfect contentment.

Again it was Imo who devised this method of cleaning grain. She must have been something of a Prometheus among her fellows.

The Ladder from Lemur to Thinking Organism

It is a fascinating undertaking to divide the human phenomenon into a cluster of separate basic traits and to attempt to pursue each of these lines back to prehuman roots, back to anthropoid apes, monkeys, and prosimians. To be sure, we have still a long way to go before we can trace the whole pattern of man's evolution. Social behavior in the wild, intelligence, capacity for speech, and other qualities have been systematically investigated only in a few species of primates. But the work is in full swing, and already the outlines of highly interesting prospects can be discerned.

These animals stand at the beginning of lemur evolution toward community living. The African galagos (A) and the Indonesian tarsiers (B) are still solitaries. The Madagascan sifakas (C) and ring-tailed lemurs (D) live in small, simply organized bands.

What is our present view of the evolutionary lines of social behavior and intelligence over millions of years?

All apes and monkeys are without exception—like man—distinctly social creatures. The higher lemurs also live in social groups, although of a simpler structure. The lower prosimians such as the galagos (bush babies) and tarsiers, however, are solitary. Or rather, they live in pairs. Thus, on the ascending scale of the primates, we have all social forms, from the simplest of units to the most variegated types of community living, laid out before us in a kind of living history of the planet.

In the course of an eleven-month expedition to Madagascar sponsored by the New York Zoological Society, Dr. Alison Jolly[16] leafed through this history. She studied the social structures of various species of lemurs and compared these with monkeys higher on the evolutionary scale. Some of her central questions were: What forces are involved in the transition from solitariness to living in communities? What is the part played by sexuality and intelligence?

Hitherto, two opposite hypotheses have been widely propagated. One school of thought maintains that even in its early stages human society was a purely rational institution, a hunting and protective alliance, and hence a product of intelligence. But there is another view, that every human and animal social order holds together solely on the basis of sexual ties. Desmond Morris[17] predicates that the 'naked ape' originated solely from the evolution of this instinct. Let us look to the facts.

The social organization of two species of Madagascan lemurs, the ring-tailed lemurs and the sifakas, differs in one fundamental respect from the communities of lions, beavers, hoofed animals, and other mammals that do not belong to the order of primates. For these expel all male and some of the female offspring out of the social group as soon as they reach the age of 'adolescence'. In so doing they cut off the possibility of further evolution toward any higher form of community life.

Apes and lemurs, on the other hand, do not drive the maturing young out of their association. And this permits a many-layered social group containing all ages of both sexes to develop. Here is a tremendous evolutionary advance, without which there would certainly be no human race today.

When such an association of multiple families grows too large from time to time—but only then—it splits up into two smaller groups of similar composition.

Let us take a closer look at the sifakas. These denizens of the jungle, about sixteen inches long and usually brightly colored, with extremely long and muscular hind legs which permit them to take leaps of thirty feet and more through the treetops, are generally models of peaceableness. In troops of up to ten members, they swing silently, one behind the other, from tree to tree. They eat only vegetarian food, in silence, without disputes.

There is no order of rank among the members of these little social groups. In this early phase of socialization rank has not yet been 'invented', as it were. Any member of the group can be the leader as they swing single file through the trees. When sifakas wrestle with one another, none tries to 'win'; their mock fights are exercise rather than conflicts or competitions. For the most part these lemurs merely grip their partner's feet and turn them like bicycle pedals. If one of them tires of this game, he simply jumps away and the other goes calmly back to feeding. What would there be to quarrel about anyhow? In such a band there is absolutely nothing that would be worth disputing.

Even 'territorial defense' is more of a ritual than an aggressive act among the sifakas. Every troop claims a jungle territory of a fourth of a square kilometer. The boundaries in the treetops are defined by scent marks. If neighbors make so bold as to disregard these limits, there is a 'battle', but it seems more like a chess game.

In these duels in the crowns of jungle trees, the victor is the one who succeeds in leaping on his opponent's back and clinging fast. Then he has all his weapons free for the fight, whereas the other must cling to a branch. This means that each opponent knows, before leaping or before being leaped at, what course the encounter will take. Consequently, there is no need for the real fight. All that counts is the initial position.

On the border between the territories of two troops of sifakas, there is a series of 'strategic' branches. The sifaka who approaches so close to an opponent that he could reach him with one leap, but crouches so much higher in the tree that the other cannot leap directly at him, has therefore checkmated his rival. The latter quickly beats a retreat, with the happy result that blood is almost never spilled. Since the lemurs make use of all sorts of refinements, diversionary maneuvers, pincer attacks and the sudden deployment of hidden 'reserves' in these mock battles, the comparison with a chess game is actually quite fitting.

Only during those two weeks in the year when the sifakas are able to mate is the pacific quality of the small band endangered. There are no lasting marriages; possession of or power over the females is not regulated by order of rank. Therefore the jealous males claw in blind fury at one another. Serious wounds are inflicted in the course of these mating battles, though Dr. Jolly has never observed a killing among lemurs.

From this we may well conclude that the sexual instinct cannot be the uniting force in such an animal society. On the contrary, during the short period in which the sexual instinct reigns, the cohesion of the troop is almost destroyed. It almost seems as if each of the animals would prefer to go his own way and live a solitary life like the lower lemurs, the galagos and tarsiers.

In this connection we must recall that Helga Fischer (*see* Chapter 3, note 9), in her detailed motivation analyses at the Max Planck Institute for Behavioral Physiology, has demonstrated the existence of an independent associative instinct (see page 64). It seems beyond question that this instinct is a biological reality.

The force of this instinct becomes evident in the lemurs, who re-discover their sense of community after mating. The newborn lemurs, who come into the world in July, are a highly effective 'social cement'.

All the members of the troop, whether females, males, or adolescents, are equally infatuated with every baby. They throng around the tiny creatures to delouse them. Incidentally, the lemurs do not pick lice with their hands, as monkeys do, but with their teeth. They use their teeth like combs, moreover, running them through each other's fur to remove the lice.

At the beginning the mother lashes out against all the 'visitors' who try to share her child. They are permitted to watch only from some distance away. Thwarted in their strong urge to delouse the baby, they delouse each other. This act of mutual grooming gradually restores the old friendship within a troop of sifakas.

Observations such as these lead to the following conclusion: the force for social cohesiveness in these animal societies is based on the associative instinct. Intelligence plays no part in the affair—at least not at an early stage of social evolution.

The small and simple societies of the higher lemurs arise by grace of an associative instinct which is motivating the animals, not on rational grounds. However, the theory of rational motivation would gain plausibility if it could be demonstrated that the lower on the social scale monkeys and lemurs stand, the less intelligent they are. And in fact Thorsten Kapune[18] of the Zoological Institute of Münster University has ascertained that the intelligence of ring-tailed lemurs, as measured by their capacity to form a 'concept of value', is considerably less than that of the higher monkeys, the capuchins and rhesus monkeys.

But what is meant by 'intelligence' in this connection? The student of animal behavior defines intelligence as the capacity to learn something. In contrast to instinctual behavior, intelligence enables animals to have experiences and later to apply them meaningfully. We are already acquainted with learning processes of an extremely primitive sort among mollusks and coelenterates, and in general among all creatures possessing a nervous system, no matter how simple. In this sense even the earthworm possesses a tiny quantum of intelligence.

Many zoologists are fond of investigating the intelligence of animals by means of tests of their own devising, the results of which are then compared with results obtained from tests of other species. The test subjects must, for example, learn to recognize several shapes, or follow paths in mazes. Or else they must earn their feed by obtaining it in ingenious ways. To that end they are made to manipulate tools, boxes, poles, ropes, or hooks.

Dr. Alison Jolly protests against such methods. She contends that no general conclusion can be drawn from artificial tests of this sort. For example, is a young man who fails Latin utterly stupid? Perhaps he manifests unusual intelligence in some other field. But anyone who tests him only in Latin will never discover that. A mouse may be stupid at differentiating flower patterns, but will do astoundingly well at learning paths by heart.

It is the same with monkeys. The American zoologist points out that intelligence in dealing with objects is the forte of men, not monkeys. If we want to understand the whole phenomenon of intelligence, we must in all fairness distinguish among various types of intelligence.

The three pillars of intelligence are fairly evident. They are respectively: enemy–prey intelligence, social intelligence, and tool intelligence. The latter can be considered the beginning of abstract intelligence, in other words, of thinking.

The first stage in the evolution of animals was enemy–prey intelligence. The eternal struggle between predators and their prey over millions of centuries has led to a cumulative increase of intelligence on both sides. Amphibians are more stupid than the evolutionary younger reptiles, and the latter are in turn less intelligent than birds and mammals.[19]

The mechanism governing the evolution of intelligence works more rapidly under conditions of numerous species and sharp competition— that is, on the large continents—than it does where there is a smaller assortment of animals, as in Australia or Madagascar. Consequently, the evolution of life in those regions has lagged by millions of years, except for 'imported' animals such as man. In Madagascar the evolution of primates has remained at the stage of lemurs. And in Australia evolution has actually taken an entirely different direction, toward marsupials, which viewed as a whole are far inferior in intelligence to mammals.

There is no direct connection between these facts and social intelligence. Among the solitary animals, social intelligence is practically zero. Every member of the species, for example among tigers or polar bears, is an enemy, except for the sexual partner in the brief mating season, and the animals' own offspring as long as they are young. The rituals of courtship and care of offspring among many solitary animals are entirely governed by instincts. Nevertheless, the first signs of social intelligence can make their appearance among these animals.

With the sifaka lemurs social intelligence is, as we have seen, not very highly developed. On the other hand, another lemur group shows distinct advances. The ring-tailed lemurs—about the size of foxes and a favorite zoo attraction because of their large, bushy, black-and-white-ringed tails—fight to establish a linear order of rank in their societies. Who is allowed to approach whom and the principle of privilege are strictly regulated in their community.

But the ability to handle lemur etiquette, to take correct measure of the partner's personality, to show psychological shrewdness, involves mental capacities that should not be underestimated—involves, in short, social intelligence.

Yet Dr. Alison Jolly was never able to observe 'protected threatening' among ring-tailed lemurs. This is a type of effrontery developed to perfection among lesser apes such as baboons, and monkeys such as the macaques. Low-ranking members of the troop, or juveniles, will resort to this trick when they have been oppressed by a sulky old male and want to retaliate. The youngster will take his stand directly behind the all-powerful boss and from that vantage point will make fun of his enemy with all sorts of mute but highly provoking gestures. The grumpy older animal is totally dismayed. He has to put up with the insults, for if he launched an attack, snarled or made counterthreats, the boss, unaware of the trickster behind him, would assume that he himself was being challenged. Then there would be hell to pay. Can human pranksters be any more cunning?

To all appearances the lemurs are not capable of such acts of higher social intelligence. The American zoologist therefore concludes that an animal society can arise without the exercise of social intelligence. But community life will stimulate the development of this type of intelligence and will thus in turn lead to more complex forms of community living. All this represents a form of reciprocal evolution whose level is constantly rising and which operates a good deal faster than the old enemy-prey mechanism.

Without this accelerating factor in evolution, no such paragon of intelligence as man could ever have arisen.

A paper by Stanford Professor Albert Bandura[20] throws a great light on this matter. There are human children who when considered by themselves alone give the impression of being highly intelligent. But in conjunction with other children they are complete failures. They lack only one special type of intelligence: the social type. This does not mean that they are in the least mentally deficient. As a rule the children who display this kind of abnormal behavior are those who grew up in too great solitude before reaching school age. The American psychologist has developed methods for helping such children make up what they have missed in 'social learning'.

The next higher stage of intelligence, tool intelligence, poses considerably more complex riddles. To be sure, in the prison conditions of zoos and laboratories some anthropoid apes perform amazing intellectual feats in handling sticks, levers, ropes, and boxes. These abilities are somewhat more limited among the lesser apes, such as baboons, and among macaques and capuchin monkeys. But even among the lemurs the first glimmerings of such intelligence may be observed. In the wild, however, none of these slumbering talents in the animals come into play. No ape, monkey, or lemur at liberty has ever been seen employing tools. The sole exception is our closest relative in the animal realm, the chimpanzee.

Why do these animals apply their tool intelligence only at the instigation or under the compulsion of human beings, but never voluntarily in their natural territories?

Dr. Jolly suggests the following hypothesis: Among these animals in the wild, will and intelligence are so heavily dominated by the search for food, fear of enemies, and social tensions that 'diversion' to lifeless objects cannot occur. This decisive step is only taken when the enemy-prey intelligence has developed to such a point that it permits a degree of emancipation from the elementary cares of living, and when the social intelligence under satisfactory conditions supports the enemy-prey intelligence.

Up to that point the tool intelligence exists only in latent form as a kind of 'surplus product' of the social intelligence, a predisposition, a promise of future evolution.

Tool intelligence requires abstract thinking. A chimpanzee who decides to prepare a stick to be used in probing a termitary must complete the whole process in his mind beforehand, with insight into all the steps involved. Anthropoid apes in laboratory experiments who must handle tools in order to obtain bananas usually approach their task with great enthusiasm. If they do not succeed at once because of inadequate experience, they withdraw somewhat, assume a meditative posture, then try another method, obviously one they have thought out.

Thinking, says Freud, is a form of testing activity. Among apes it is of course a thinking without words. The chimpanzee considers something in nameless concepts. That must be far more difficult than thinking with the aid of a verbal language. When we appreciate that fact, we see to what far reaches thinking with the aid of language can attain—not to speak of thinking with the aid of writing.

At the present stage of evolution, the existence of these separate types of intelligence has had one dire consequence for man. Tool intelligence and abstract thinking have led him to tremendous advances in the command of techniques. But man fails to solve the problems of social coexistence—for the simple reason that his social intelligence has scarcely evolved beyond the level of the apes.

9. The Dawn of Man

From Bite to Smile

The humorist Erich Kästner once wrote: 'Columbus discovered America, and the Greek philosopher Aristotle discovered the sole difference between man and all animals. He observed that only man can laugh, and therefore called him the laughing animal.'

With all due respect to Kästner and Aristotle, modern behavioral science must revise that statement. For there are animals that can smile, grin, and even laugh heartily. For example, all apes and monkeys and almost all lemurs smile. At each stage of evolution, however, smiling has a somewhat different meaning. In fact, an evolutionary line can be traced from rudimentary beginnings to the full-throated laugh.

Among the lemurs, tarsiers, indri, lorises, galagos, and pottos, the appearance of the face is relatively unchanging. These animals almost entirely lack the facial musculature needed for expression. All they can do is bare their teeth before biting, or when trying to frighten an enemy. But baring the teeth is the first step from which smiling has evolved. Smiling and laughing are basically ritualized biting, says Dr. N. Bolwig.[1]

Among the lemurs, those who live in some form of community show greater variability of facial expression than the solitaries. When the brown lemur (*Lemur fulvus*) must make arrangements with members of its band, there will be a showing of teeth much more distinct than among the dwarf or mouse lemurs, both of which are solitaries. Other mammals show a similar tendency. The solitary bear has a totally fixed facial expression. There is no telling what is going on in his mind or what he intends to do. But a man can fairly well interpret the play of expression in the faces of sociable dogs and wolves.[2] Members of the pack are no doubt able to read these expressions perfectly.

Not the slightest expression stirs in the faces of these lemurs. The West African potto (A), the Ceylonese loris (B) and the Madagascan indri (C) do not even have the muscles necessary for facial expression.

Among animals that can change their facial expression, each mien has its special meaning, just as it does among human beings. Thus a social lemur can threaten a rival by opening its mouth wide but keeping the teeth covered by the lips. This amounts to saying: If you don't leave immediately, I'll let you feel my teeth!

There are other nuances. If the lemur just barely opens its mouth, drawing back the corners and gnashing its teeth—oddly enough this means just the opposite of a threat. It is an assurance that there is not the slightest intention of biting, as if the animal were saying: See, I have teeth that I could use to bite, but I won't. Here is an evolutionary advance: the first trace of a smile, a demonstration of friendship which is not to be confounded with weakness.

Grinning is closely akin to smiling. Dr. Richard J. Andrew[3, 4] calls the grin a smile in which the amicable mood is lacking. That definition applies to men as well as monkeys.

The smile of friendliness has been further refined by the lemurs and the apes and Old World monkeys of Africa and Asia; it has become ritualized into a gesture of humility and appeasement. If, for example, a weak baboon wants to be spared a beating from a stronger one, or after fighting wants to escape further punishment from the winner, he must throw himself to the ground like a Moslem about to pray, turn his upthrust backside to his adversary, at the same time looking over his shoulder, grinning broadly, and audibly smacking his lips. In baboon language that means approximately: Don't do anything to me! I promise to be good from now on.

The gesture of appeasement invariably checks the hostile attack. Even human beings can save themselves from baboon bites by performing this act (see page 14). But without the grin the gesture is only half effective. It is significant that there is no friendliness at all in this grin, but a good deal of fear.

A grin likewise appears on the faces of animals when high and shrill cries of fear are uttered. Dr. Andrew found that it accompanied the clicking and prattling sounds of lemurs, the piercing screams of baboons, and the screeching of quarreling monkey babies. In such cases grinning serves as a form of defensive threatening. One root of smiling, then, is to be found in fear.

Psychologically, that fact is highly illuminating for us human beings. Children often smile unexpectedly when they are being reproved by teachers or parents. Unfortunately many adults misinterpret this reaction as the height of impudence. In reality the children's 'smile of embarrassment' is a sign that they feel deeply disturbed. The psychic upheaval triggers a behavioral relic of primordial times: the humility gesture of our apelike ancestors, which is rooted in their instinctual patterns.

On the other hand, the mimicry of smiling can never entirely deny its aggressive origins, the threat of biting. It is equivalent to saying: It would be nice to have flesh between these teeth, but I had better content myself with air rather than risk being bitten back. There is a liberating smile or laughter which eases tense and aggression-fraught situations. But there is also a kind of grinning which can provoke assault. In so-called gentlemanly circles people cultivate the art of throwing the cruelest insults into the faces of opponents in well-chosen words, accompanied by a cold smile.

Like human beings, macaques can feel that they are being laughed at. These monkeys practice duels of smiling. But such smiles often abruptly turn to the *urr-urr* sounds of aggression.

With lemurs, macaques, and baboons, slight shades of expression distinguish the smile of humility from the grimace of naked fear. Before a dangerous leap, or when they are afraid of falling from the tree, these animals grin. Monkey mothers and babies who have lost touch with one another likewise smile even while they are uttering heartrending wails of misery.

The smile of greeting, which has formalized into a mere ritual, no longer contains anything but faint overtones of anxiety. In human as in ape and monkey society, custom demands that the lower in rank greet the higher in rank first at every encounter, by smiling respectfully. Greetings, then, are originally nothing but a routine gesture of submissiveness.

Animals, however, must pay attention to another important aspect of the matter. The lower in rank must never look the other directly in the eyes in the course of a greeting. Such direct looks bespeak aggressiveness and would immediately convert a smile of greeting into a provocation. For that reason, when apes greet they glance just past the face of the one they are greeting, or else they look down at the ground, like well-bred Orientals.

Gradually, from this behavioral compound of elements of anxiety, appeal for safety, and wish to appease, the sign of friendliness and friendship has evolved.

We have not yet spoken of the form of expression which we human beings esteem so highly: smiling and laughing for joy. It would seem difficult to trace these modes of expression to any animal source. The fact remains, however, that anthropoid apes can utter full-throated laughter from sheer pleasure. Chimpanzees, orangutans, and gorillas possess a facial musculature very similar to man's, and they use it as we do in similar situations. These animals laugh cheerfully when they tickle one another in the armpit, on the neck or the sole of the foot while grooming, when two old friends meet each other in the jungle after a long separation, or when they have found a special dainty while searching for food.

Chimpanzees have a distinct sense of humor. If one chimp means to tease another, the corners of his mouth twitch even before he has acted, Dr. Bolwig has observed. And if the joke succeeds, the ape laughs heartily. Young chimpanzees will play peek-a-boo with the same delight as nine-month-old human babies. When the familiar face is first hidden, then comes into view again, the baby chimp giggles excitedly.

Saimiris can chirp with pleasure when they receive food, or when a person whom they know well comes into the room. Peter Winter (see Chapter 7, note 3) translates the birdlike twittering as: Look, here's something pleasant! As they make the sound, the monkeys contort their faces into a smile. In these gestures we may probably discern the origins of smiling for joy.

Joy in animals is the reaction to an unexpected event which arouses pleasant feelings. Richard Andrew[3] suggests that grinning was originally a reaction to frightening stimuli, whereas smiling was the response to minor and pleasant changes in stimuli. He finds support for this view in the fact that babies respond by smiling when they are tickled, or when someone plays peek-a-boo with them. Even among adults, he holds, laughter at jokes arises from the effect of surprise.

To be sure, even among chimpanzees pleasure and pain are closely related. If the chimpanzee opens his mouth and shows his lower teeth, while the upper teeth remain concealed by the lip, he is laughing. But if he also bares the upper teeth, as a laughing human being does, he is signifying extreme fury or great pain. Unfortunately inexperienced attendants in small private zoos or travelling circuses misinterpret these expressions, and think the animal is in good humor when he is in the worst of humors. Then the poor animal is saddled with the reputation of being treacherous, and receives many a beating.

Such misunderstandings can even lead to unpleasant scenes among closely related lesser apes. Thus the mandrill[5] will signify his friendly intentions by gnashing his teeth and giving a cry of appeasement. To his cousin the baboon the same combination means the exact opposite: a serious threat. Consequently, if the two animals are locked up in the same cage they get along as badly as dog and cat. Wise keepers can however, gradually let them grow used to one another until each has learned the other's 'foreign language'.

The ape with the most human laugh is the orangutan. It takes no special knowledge to understand his facial expression. He is master of all nuances from smiling astonishment to spiteful gloating to hearty laughter over a joke. The long evolution of facial expression from biting to laughter has reached its height in this animal.

There is another evolutionary line of far greater significance that branches off from this one: the road to human speech. The old theory was that the capacity to articulate sounds suddenly arose with the evolution of man and, moreover, at a point in time at which the evolution of the mind had correspondingly made great progress. Dr. Andrew, however, considers a leap from mute nothingness to full language highly improbable. He believes that preliminary stages of the capacity for speech can be discerned in the form of primitive facial and tongue movements.

The baboons offer some support for this view, since their gestures of appeasement and greetings are accompanied by a broad grin and a smacking sound. Deep grunting may also accompany the smacking, the animals modulating the sounds by lip and tongue movements. Sonic spectrographic recordings have fully verified such changes of tone. Thus the baboons have taken the first steps toward forming different sounds in the mouth.

Baboons also utter grunting noises when they are engaged in tender 'love talk' in pairs. These sounds have not yet been analyzed closely, however. Possibly they will be found to occupy a level similar to the first frail sounds human babies produce when suckling or when kissing open-mouthed.

But the question of speech needs to be considered in a broader context, that of evolution in general.

From Pantomime to Verbal Language

It was with such larger ambitions that Dr. Adriaan Kortlandt[6] made several expeditions to the Congo jungle and the savanna of Guinea to study the language of chimpanzees. His findings were highly interesting.

Communication among chimpanzees in the wild consists chiefly of 'talking with facial expressions and with the hands'. Holding up the hand, that official gesture of the policeman, means exactly the same thing: 'Stop!' These apes likewise signal 'Come here' or 'walk quickly past me', with gestures that are amazingly human. The hand held outstretched in a begging gesture (see page 18) signifies a greeting, a plea, or a recommendation to a fellow chimp to calm down. The reciprocal greeting or the gesture of accord consists in holding out the hand reversed, that is with the palm down. In such a gesture the fingertips of the two animals may touch. But the gesture can also be well understood at a distance.

Two good friends occasionally greet each other by a 'handshake', if they have not the time or the desire for an embrace. Even when they are in a great hurry, or fleeing, they perform a sketchy version of this greeting. In Dr. Kortlandt's experiment with the stuffed leopard the chimpanzees 'encouraged' one another by using this gesture and kissing hands.

Chimpanzees will sometimes warn the fellows of their band of a distant danger, such as a snake, by using their outstretched arm and pointing with their index fingers. That is something altogether extraordinary in the animal world, for aside from bees, geese, and ducks, no other animals can point a direction to members of his species.

Dr. Kortlandt has also seen a number of gestures whose meanings were less plain. Once an ape used a signal consisting of raising his arm slantwise three times in succession, as if intending a greeting, but with clenched fist. There were times when the animals were surely conversing, as when after gathering food they all sat together having their meal, and communicated with facial expressions and gestures, like deaf-mutes.

Once the scientist saw a male holding an open papaya up to his companion's eyes, presumably because a grub was crawling inside it or something else was amiss. 'These animals, to be sure, are not capable of intellectual effusions,' Dr. Kortlandt says. 'But they can grumble endlessly about trivialities.'

We know that chimpanzees are even capable of reading. This does not mean that they can pronounce or decipher the letters of an alphabet; but in captivity they can learn to recognize reduced, schematic line drawings of familiar objects. That is certainly the first step toward understanding a kind of symbolic language.

What is more, some talented chimpanzees in the London Zoo spontaneously—that is, without training—perform pantomimes. Like Marcel Marceau they eat nonexistent meals, making all the proper movements, employing plates and silverware that exist only in their imaginations. Such perfect mimicking reveals that chimpanzees are quite capable of forming 'pictures' or 'conceptions' of unseen things and of communicating these to their fellows.

'That is,' Dr. Kortlandt concludes, 'they have created the very essence of genuine language, including conceptualization and symbolization, though in a non-verbal way. Thus it would seem that in the course of protohominid and early hominid evolution, non-verbal pantomime "conversation" about the unseen came first, verbal talking much later.'

There is one difficulty with this hypothesis: the riddle of babbling among baby chimpanzees. For chimpanzee babies babble just like human babies. But whereas in human babies babbling has a purpose—to exercise the sensory and motor control mechanisms that are necessary for speech—it seems to lead to nothing in the chimpanzee babies. At the age of four months they abruptly stop all sound production.

At this age, apparently, an inhibition matures in the chimpanzee baby—a useful one, which represses the unnecessary chatter which would spell trouble in the dangerous conditions prevailing in the jungle. This premise is supported by an observation made by Dr. C. Hayes[7] when she attempted to introduce human sounds into her play with a young chimpanzee. The baby started in fear and became so nervous that it would never again play that particular game.

In the jungle, chimpanzee mothers and babies communicate almost exclusively by looks and gestures—for a leopard may be lurking behind every bush. But possibly the ancestors of the chimpanzees, who lived in the open savannas where places of ambush were fewer, communicated with the aid of sounds similar to words. In general only the mothers and babies are so silent; grown male chimpanzees make a tremendous amount of highly varied noise—when they feel secure.

The inhibition to the production of sounds may have arisen only after the animals were driven back into the jungles by the ancestors of man. Consequently, the babbling of chimpanzee babies may represent a vestige of early prehominid speech which degenerated once again in the course of chimpanzee evolution.

Thus the failure of chimpanzees in all attempts at speech may be caused less by inadequate intelligence or poor ability to articulate than by an inborn inhibition which may be roughly compared with the inhibition that makes a stammerer. For after all, we should really expect an anthropoid ape to perform at least as well as a parrot in regard to speech.

In captivity, as a matter of fact, the vocalization of chimpanzees[8] is by no means negligible. They can pronounce the vowels from a to u, though they prefer o and u. They express delight by an *oh* uttered several times, the pitch rising with increasing excitement. They utter warnings by using a low short *eh* and grief by low *u* sounds.

The vowels can also be combined into phonetic groups with the consonants k, g, ng, w, wh, and h. *Gak* means: 'I have found food', *kuoh*: 'I am very hungry'. *Gho* is spoken in greeting friends, and *ky-ah* is a cry of pain. *Vts* is whispered when the animals are picking lice; *ah-oh-ah* expresses moderate concern. The chimpanzee's *ah-ee* and *oo-ee* correspond to our human 'ouch'.

A chimpanzee can learn to obey as many as fifty different verbal commands. In other words, its understanding of sounds and memory for them is quite respectable. But it has difficulty in speaking itself. The most that Dr. Hayes was able to teach her hairy pupil after many years of toilsome effort were the words *mama, papa, cup,* and *up.* And even these words the ape could manage only with much grunting, gurgling and production of saliva. The Gardners,[9] who taught a young female chimpanzee sign language, were considerably more successful. When this chimp, Washoe by name, who has been at the Gardners' laboratory in Reno since she was a year old, has finished eating, she brushes her index finger over her incisors to indicate that it is now time to brush her teeth. But if she wants more to eat, she places the fingertips of both hands together. If all she wants is a few sweets for dessert, she moves the tip of her tongue back and forth, meanwhile touching it with her index and middle fingers.

By the time she was four, Washoe had attained a vocabulary of more than sixty words—and this, moreover, in standard deaf-mute sign language. The scientists made only a few adjustments in the sign language in conformity with the nature and talents of chimpanzees.

When the ape wants to say 'please', she runs the palm of her hand over her chest. A similar movement with closed hand means, 'I'm sorry'. Both index fingers outstretched and moved crosswise signifies, 'It hurts'. As for 'hurry up', and 'please give me', these are expressed naturally, with the same gestures wild chimpanzees would use.

But size of the vocabulary is not the whole story. It was rather the way Washoe employed this dumbshow that overturned previous scientific notions of the limited ability of animals to communicate. The contrast with parrots was striking. For Washoe correctly transferred expressions whose meanings she had learned in connection with specific objects to other things of the same sort. For example, Allan and Beatrice Gardner had taught her to use the sign for 'open' when she wanted the door of her room opened. Without any further instruction Washoe used this sign to ask that other doors be opened, or if she wanted someone to open the refrigerator or a box.

Having learned the symbol for flower, the chimpanzee applied it to all flowers, large and small, single and in bouquets. Obviously she had grasped the botanical idea of flower. Washoe uses the signal for dog when she sees a dog, but also when she merely hears one, although no one has taught her that. In other words, she can translate things she does not see into optical language.

What is more, after twenty months of instruction in sign language Washoe began combining two and even three of the learned symbols into self-invented meaningful sentences: 'Please—give me—a scratch,' or 'Me—go out' or 'Please—open—quickly,' or 'You—drink.'

This ability to join single words deliberately and meaningfully, so that complicated communications and even simple sentences are produced, has hitherto been regarded by linguists as an exclusively human trait. There, it was thought, the line between man and animal could be drawn clearly and unequivocally. This view, too, can no longer be maintained, at least not in the area of gesture language. Whether animals can do anything similar with sound language has not yet been discovered—not yet, we must in all caution emphasize.

In our search for the primeval origins of human speech the investigation of chimpanzees has so far led us nowhere. Let us therefore eavesdrop on members of those primate species which are by nature less taciturn.

The saimiris, for example, have a signal system that employs a great variety of sounds. They can produce squeaks, twitters, cackles, barks, hisses, squalls, chatters, and screams—essentially a larger gamut of sound than man employs in his languages. Here we stand at a fork in the road of evolution. Quite conceivably the path to speech might have taken the direction of a multiplication of sound effects, from grunts all the way to whistles. It would be amusing to imagine what conversation would sound like then!

Evolution took another course, however. In the higher apes, most of the sound effects employed by the saimiris have been abandoned. To be sure, an ascending trend does not necessarily lead to the crown of creation. In the evolutionary line of monkeys and apes the one element in the phonetic repertoire of the saimiris which has remained unchanged despite the passage of seventy million years is screaming. Consequently, screams are employed with the same meaning by all lemurs, monkeys, apes, and men: as the expression of intense excitement.

According to Dr. Peter Winter (*see* Chapter 7, note 3) the squeaking sounds are of much greater evolutionary importance. Within this category the saimiris produce four variations. The 'squawk' is always uttered when one of the animals has lost sight of its companions. They answer in the same manner until the strayed saimiri has found them again. On the other hand the 'contact squeak' is lower and shorter. It is used by the monkeys when they can still see one another, but the farther apart they are, the more frequently they use it.

The 'alarm squeak' sounds like a badly overwound alarm clock. It is used to report every fast-moving object such as a jaguar. When this cry sounds, every monkey dashes for a hiding place and remains there motionless.

In addition there is 'cheeping'. This is the already mentioned sound of friendliness. It means something like: I don't want to do you any harm; I only want to play. Monkeys wrestling in fun keep up a continual cheeping. Otherwise the game would at once turn serious.

In contrast to the saimiris' wide repertoire of phonetic types, the vocal spectrum of macaques and anthropoid apes, which stand higher on the evolutionary scale, seems most meager. They have lost the twittering sounds that express creature comfort. Instead, they exemplify a new evolutionary trend which has gradually led to human language: while the phonetic types among these animals has been much reduced, this is more than offset by the increased variability within the remaining phonetic types.

In addition, the higher monkeys and apes have developed the ability to combine single phonetic elements. In so doing they can express emotional reactions with considerably greater variety and subtlety. Here it would seem that evolution took a two-laned course, one in the direction of greater pantomimic expressiveness, the other in the differentiation and combination of sounds.

The latter procedure can be carried fairly far in, for example, species of monkeys and lesser apes which live in large troops and are seldom exposed to acute danger from predators. Examples are the red-faced

Japanese macaques and the baboons. The latter, because of their well-organized fighting abilities, need not be much afraid of leopards in their forested plains.

Nevertheless, red-faced macaques and baboons, though fairly expressive at pantomime, have not done very much with sounds. Both forms of expression are innate in them and are used largely to communicate moods. Aside from certain peculiarities, the pantomime is 'international' within the species, just as human pantomime takes much the same form among Europeans, Eskimos, and South Sea Islanders.

The distinctive element in human language, however, is precisely the fact that it can be invented, reshaped at will, and learned.

Otto Jespersen[10] reports an astonishing example of this ability. In Denmark the Youth Board had to assume the tutelage of two children who had grown up solely in the care of a deaf-mute grandmother, completely neglected and isolated from all other human beings. They knew no other people and no one had ever taught them to speak. Nevertheless the children constantly chattered with one another, using a great many words. Out of nothing they had developed their own language. No one else could understand it, of course, because it bore no resemblance to Danish.

Are there any indications, no matter how slight, that monkeys are moving in this direction? The scientists believe there are.

Professor M. Miyadi[11] bases his opinions on years of observing the Japanese red-faced macaques. Here, for example, is how a male monkey makes a proposal: With prancing steps he approaches the female, protrudes his lips and smacks them with relish. Thereupon the two form a pair and leave the troop for a short time. The rhythmic lip movements are a sign of warm feelings, and are also common among young females. This mode of communication is wholly personal, directed from one individual to the other, and serves a function similar to that of human conversation.

There are many indications that along with the innate and instantly comprehensible sound signals for alarm, departure, staying, threatening, fear, and invitation to mating there are more subtle 'subjects of conversation' among these monkeys. Professor Miyadi noted very low-voiced utterances which reminded him of people murmuring to one another.

Dr. Andrew[12] is even more certain that the primates are reaching out toward the creation of language. He describes how baboons imitate grunts when someone grunts to them, and how they grunt to one another in sociable situations. He concludes: 'In this way, the first requisite for the development of vocal language, namely the learning of a particular vocalization pattern from a fellow in association with a particular situation, would have been achieved.'

But the phenomenon of human speech must also be examined from another side, in terms of the physiology of the brain. Is there a speech center? Where is it situated? How big is it and how does it work?

Nature's Great Puzzle

There are people living among us who have undergone a remarkable form of surgery: physicians have separated the two hemispheres of their brains. Each part of their brain works independently, as if two altogether different and entirely complete brains were located in one cranium. Careful studies in 1967 have shown that such patients experience two distinct realms of consciousness. Probably they even have two sets of thought and two sets of feelings at the same time.

The actual purpose of these operations was to cure severe forms of epilepsy. But they also yielded some surprising insights into the nature of consciousness and the talent for language and speaking.

The exciting story of these researches began some fifteen years ago. Professors Ronald E. Myers and R. W. Sperry[13] at the University of Chicago had posed the question of what would happen if the nerve connections between the right and left hemispheres of the brain were severed by surgery. Suppose this experiment were carried out on cats—would the cats live, and how? At first sight, the animals operated on seemed to behave exactly as they had done in the past. But an ingenious test brought remarkable things to light.

The scientists taught one of their cats to press a lever with her right forepaw as soon as a black cross appeared. The animal quickly learned to perform this task. But she was altogether at a loss when she had to do the same thing with her left paw, and had to start learning the operation from the beginning.

What explained this? The muscles of the left side of the body are directed exclusively by the right hemisphere of the brain, and the muscles of the right side by the left hemisphere. In normal organisms there is a constant exchange of information between these two parts of the brain. But the connection had been severed in the cat. Consequently her left paw could not possibly know what her right paw was doing. With the brain divided, each half of the body had to learn anew things that the other half could do perfectly well.

Aside from this, the animals showed no detrimental effects. That encouraged a team of brain surgeons, J. E. Bogen, E. D. Fisher, and P. J. Vogel[14] at the Medical Institute of California to attempt the same operation on human beings in 1961. Their subjects were patients suffering from severe, uncontrollable, and hitherto incurable epilepsy. Their hope that they would be able to confine epileptic attacks to one side of the body was far surpassed by the result: the attacks, including the one-sided ones, ceased almost completely after the severing of the two hemispheres of the brain.

The psychobiologist Dr. Michael S. Gazzaniga[15] carefully studied the patients whose operations had been so unexpectedly great a success. There seemed to be no change in the personality, temperament, and intelligence of the former epileptics. One patient remarked only that now and then he had 'split headaches'.

Soon, however, some oddities in the daily life of these patients turned up. If they bumped into a chair or a table with the left sides of their bodies, they did not notice. If a coin or a spoon were placed in their left hands, without their seeing it, they obstinately denied that they were holding anything.

That evidently had some meaning, and Dr. Gazzaniga thereupon decided to undertake a series of probing experiments. He had his patients stare fixedly at the center of a screen, for example, and flashed the word 'heart' on the screen for a tenth of a second. The projector was so arranged that the letters 'he' were to the left of center and the letters 'art' to the right.

A number of cards bearing different words were spread out for the patients. They were asked to pick out with their left hands the card showing the word they had just seen on the screen. All the patients with split brains pointed to the word 'he'. Then came the weird concomitant. When asked immediately afterwards to say aloud the word they had read, they unhesitatingly said 'art'.

The same letters, the same person, but two contrary opinions! What had happened?

It is known that pictures in the left half of the visual field—as both eyes see them—are transmitted only to the visual cortex in the right hemisphere of the brain and those in the right half of the visual field only to the left hemisphere.[16] In the present case the right part of the brain registered only 'he' and communicated this to the left hand, which then correctly acted on the message it had received.

But why was the subject powerless to express this observation in words? Why could he report only what was visible in the other half of the visual field? From this Dr. Gazzaniga concluded: 'Clearly, then, the patients' failure to report the right hemisphere's perception verbally was due to the fact that the speech centers of the brain are located in the left hemisphere.'

This discovery, which came as a total surprise, opens up some entirely new prospects. It had been supposed that man's 'speech center' was symmetrically divided between both parts of the brain. There is nothing visible in the structure of the brain which would suggest that our capacity for speech is located in only one hemisphere. But that is the fact; it has now been amply established.

This leads us to the principal question of the varying mental capacities of the hemispheres in the human brain. In multitudinous experiments Dr. Gazzaniga showed his patients all sorts of pictures or words in the right-hand visual field, or placed a cigarette lighter or a pair of scissors in their right hands in such a way that they could not see what they were holding. In every case the subjects were able to say what they had seen or touched. They could read writing aloud and solve

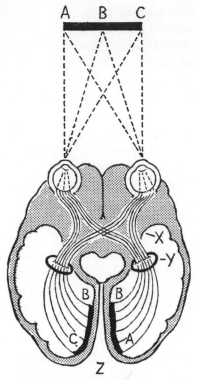

The visual cortex (Z) of the right hemisphere of the brain registers only the left half of the panorama A–B–C, while the left hemisphere of the brain interprets only the right half of the scene. This is because the optical nerve fibers follow a path from the eye across the optic chiasma (X) and through the primitive optical center (Y) to the visual cortex (Z) of the cerebrum.

problems, such as arithmetic examples. But when the scientist presented the same persons with the same things in the left side of their visual field, or in their left hands, they replied with the most nonsensical guesses, or lapsed into embarrassed silence. Sometimes they would identify a pencil as a corkscrew or an ashtray.

Does this mean that the right hemisphere of the human brain does not rise above the mental level of a feeble-minded person? To test this proposition, Dr. Gazzaniga had the subjects reply not in words, but by making signs with their left hands. Amazingly, their responses then became completely accurate. But if only seconds later they were re-quested to state in words the answers they had just given correctly, they failed. Hence it was not a matter of feeble-mindedness. It must be rather that the speech center in the separate left hemisphere of the brain knew nothing of what was going on in the other half of the brain.

It follows that if an animal cannot speak as we do, this need not be a consequence of 'feeble-mindedness'. The example of the chimpanzee who could obey fifty human verbal commands shows that the animal possesses a considerable understanding of speech. But he apparently lacks something necessary to fulfill his mental capacities: that tiny part of the brain structure which enables us to speak.

But speech is not entirely presided over by the left hemisphere of the human brain. The right hemisphere can understand spoken words very well—as with the chimpanzee. It can also comprehend writing and can command the left hand to pick out cardboard letters which make up the words it has seen, heard or felt. But to pronounce the words it has just put together is beyond its scope.

Probably the right hemisphere of the brain is also able to command the vocal organs to utter a few simple cries like 'ouch' or 'oh'. This would mean that in linguistic abilities it is on the same plane as animals. But it utterly lacks grammatical capacity—one of the foundations of human speech. The right hemisphere of the brain cannot even form the plural of a word on request.

But the matter becomes even more mysterious when we learn that both abilities—understanding language and speaking it—are equally well developed in both hemispheres of the brain in children up to the age of four years. Neurological studies have established this. Moreover, studies of linguistic development in children seem to indicate that the human capacity for grammar is innate and fully developed by the age of two or three years.

Thus the capacity for speaking is considerable in the right hemisphere of the child's brain. It begins to atrophy, however, with increasing age. Such a development is hard to understand.

On the other hand, the right hemisphere possesses characteristics that are lacking in the left. Dr. Gazzaniga asked his patients to copy pictures of a cube and of a simple blockhouse. All performed the task satisfactorily with their left hands. But they could produce only scrawls with their right hands, even though they were right-handed. Consequently the capacity for spatial drawing must be sought only in the right hemisphere of the brain.

On the other hand, the ability to distinguish between true and false is limited to the left hemisphere.

The following experiment proved remarkably significant. The scientist flashed a picture of a naked woman on the screen for a tenth of a second. When the subject perceived the picture with the left hemisphere, he laughed and said correctly that he had seen a nude. On the other hand, all those who had grasped the same picture with the right half of the brain declared that they had seen nothing. Nevertheless they giggled! Asked why they were laughing, they gave replies such as: 'I don't know . . . nothing . . . oh, what a crazy machine.'

There could hardly be a clearer demonstration that emotional reactions of this sort are just as much at home in the left hemisphere as in the right. Both hemispheres were somehow conscious of what had been seen. But only the left half of the brain was capable of putting the emotional effect into words.

Can it be asserted on the basis of such experiments that the right hemisphere of the human brain is the 'animal' hemisphere and that what is truly human is concentrated in the left half? Such a conclusion would be an excessive simplification of extremely complicated matters. Rather the experimental series as a whole should dispel some illusions about the enormous difference between man and ape.

In physical functions this difference is certainly not so great as many theorists have long considered. In many characteristics (though not in all) the differences are only a matter of degree. But the effect of relatively small physiological improvements is tremendous. We should not, however, confuse cause and effect.

Above all Michael Gazzaniga's experiments suggest that the phenomenon of speech is one of the great puzzles of nature. In the left or in the right sides of the brain various fundamental elements of this unique capacity have their seat: recognition of language, memory for language, speaking ability, language ability, grammatical ability, capacity for abstraction, writing, spatial sense, link to emotions, and certainly much else.

It is hard to believe that all these characteristics, each of which is in itself a miracle, and almost all of which appear incipiently in one animal or another, can possibly coalesce in one single organism and work harmoniously in such brilliant combinations. The very notion would be preposterous—if man did not exist!

Herbivore and Carnivore

What then distinguishes man from animal? We have just seen how difficult this problem becomes when we investigate isolated phenomena such as language, intelligence, the ability to imagine unseen things, the capacity to transmit learned material to offspring. In our present state of uncertainty about our biological position in nature, Dr. Adriaan Kortlandt[6] in 1965 suggested an interesting hypothesis. It casts the evolution of *Homo sapiens* in a new light.

The question is: To what biological processes do we owe the fact that man could arise and assume his unique place on this planet?

Dr. Kortlandt believes he has found an important clue in the observation that man is the only specialized carnivore among the higher primates. All monkeys and apes are preponderantly vegetarians. They hunt and eat animals only exceptionally, during the dry season, say, or when chance drives some prey into their arms.

Kortlandt postulated in an interview with the author: 'It is the combination of typical primate characteristics with some typical predator characteristics out of which, ultimately, something biologically so unprecedented as man arose.'

A statement of that sort seems as debatable as it is provocative. But the Dutch zoologist has provided a number of intriguing examples to support his hypothesis.

Chimpanzees possess a static view of the world. They adjust to changes in their environment, but do so reluctantly. In the zoo most chimpanzees have to be given their food in the same feeder every day, and at the same spot. If it is placed somewhere else, they react as if it were altogether unfamiliar food, and quite often refuse to eat. For meat eaters, on the other hand, changing the place of the feeder is no problem, as everyone who has a cat or dog knows. That has nothing to do with intelligence, but solely with the fact that carnivores, that is hunters, subjectively experience their environment in much more dynamic terms and can by nature adjust more easily to altered circumstances, because the animals they prey on wander about everywhere.

We suddenly perceive the implications: the chimpanzee's ability to form ideas in his mind (see page 209) in combination with the carnivore's dynamic view of the world opens the way to the creation of a mechanical technology.

Apes, as we have seen, use tools to a certain extent—stones to crack nuts, sticks to go after termites, leaves for napkins and toilet paper, heavy clubs to beat leopards. But the static nature of their view of the environment keeps them from engaging in anything like the breathtaking tempo of technological progress of which man is capable.

By breath-taking technological progress we do not mean the 'invention explosion' in our present technological age. We are rather thinking back ten or fifteen thousand years, when an important invention was made perhaps every thousand years: pottery, for instance, permitting convenient gathering, serving, and preserving of food, or the invention of the plow or the wagon wheel. And although to this day there are still 'underdeveloped' peoples who use a cow to pull a wooden stake through the ground, word of a steel plowshare not having reached them—even so the tempo of technological progress is breath-taking in comparison to the development of chimpanzee technology in the jungle.

It is no accident that in primordial times the ancestors of man invented the spear, and not the ancestors of the chimpanzee, which have never gone beyond the clubs they use in their battles against leopards.

There is a similar difference in perseverance in pursuit of a goal. As a rule nonhuman primates cannot concentrate on a single matter longer than fifteen, or at most thirty, minutes. Among the exceptions are chimpanzees fishing for termites, who are impelled to go on by the tastiness of the tiny portions, and orangutans, who move so slowly that everything takes them longer.

Carnivores, however, can lie in a bush for hours and even for days, waiting for their prey. We need only think of a pack of wolves following a caribou herd. Or let us recall the hunting lessons that lionesses give their cubs (see page 130).

This fundamental difference in disposition must lie in the fact that a plant eater with great perseverance would die of hunger as quickly as would a meat eater without that trait. A chimpanzee surrounded by papayas who had his heart set on unavailable bananas would be in a bad way. So would a lion who let his mind be distracted from his quarry.

'Since all the specifically human cultural skills and products involve persistent working towards a remote goal,' says Dr. Kortlandt, 'we may conclude that the evolution of the specifically human type of culture required the *combination* of the manual dexterity of the arboreal fruit-eater with the long-term foresight and perseverance of a highly specialized carnivore!'

We need therefore scarcely be surprised that it took so long before man arose on this globe of ours.

Nevertheless, any such combinations of traits do not suffice to explain the unusual nature of human society in contrast to all animal communities. We can very well imagine carnivorous apes or Australopithecines (prehominids) who would be equipped with all these traits and would still possess only an animal form of society. That society would never be able to produce anything even remotely resembling a human type of culture.

The fundamental thing about human society is the social, moral, and ethical code which determines what is to be done and what not done. From this code springs that which is generally called the tradition of the community and which goes far beyond the sense of belonging to a 'primal horde' characteristic of animal societies. What is more, the code is transmitted from generation to generation, so that morals and customs remain relatively constant over long ages.

It is nevertheless astonishing that both elements, the transmission of customs—that is, tradition—and a social code already exist on the animal plane. But they are separate, not combined.

Every chimpanzee must learn from its mother what is edible. She teaches it to distinguish between palatable and unpalatable fruit according to her experience. The baby chimpanzee scarcely ever touches anything that its mother has not previously held in her hands. Should it do so, the mother usually takes the object away from it.

The result is that the chimpanzee is a strict traditionalist about food. In regions where man has only recently planted papayas or corn, he will not eat these foods, which elsewhere members of his species devour greedily. Even the technique of getting termites out of their fortresses, which are as hard as concrete, has not yet been passed around to all bands of chimpanzees. Each ape community has its own fixed tradition.

But this kind of tradition is limited to one aspect of behavior, eating habits. No sign of the transmission of a social code has ever been observed in chimpanzees. A trainer can teach an ape certain habits and induce him to repress other sorts of behavior, but the primate will always break the rules as soon as the 'policeman' is out of sight and hearing.

That is not due to any failure of memory, for when the trainer returns the miscreant immediately begins screaming, from fear of punishment. Rather, the animal completely lacks the capacity for transforming learned social rules into moral commandments. The basis for such a transformation is simply not present. Apes are by nature amoral. Predatory animals that hunt in groups, on the other hand, behave quite differently. That is because such animals as lions, hyenas, and wolves must rely upon one another if their hunt is to be successful. A well-trained dog obeys the rules it has been taught whether or not the master is present.

When a dog has done something wrong, either by mistake or because the temptation was too strong, he shows clear signs of guilt feelings—whether or not his master has discovered the misdeed and in some cases even when his master chooses to be forbearing. Konrad Lorenz's dog[17] once killed a goose belonging to the household while the family was away. After Lorenz returned home he found his dog only after a lengthy search. Ever since his crime the animal had lain in a hiding place without eating or drinking. An ape would never behave that way.

On the other hand, wolves and dogs lack the chimpanzee ability to transmit learned material to their offspring. No bitch has ever been seen teaching her pups not to relieve themselves in the house or not to gnaw master's shoes. In such matters man must train every generation of dogs anew.

'Thus the unique thing in man,' says Dr. Kortlandt, 'is that he has *integrated* the mechanism of the superego with the mechanism of cultural transmittance of knowledge and skill from one generation to the next. In this way the combination of two pre-existing principles of behavior organization into one system has created something entirely novel and unprecedented.'

But that is still not all. In contrast to animals, which can already understand symbols (see page 209), man is apparently the only organism who can extend the meaning of a symbol.

Here is an example. Chimpanzees are quite able to recognize the symbolic value of coins. In numerous experiments scientists have placed slot machines filled with bananas in their cages and given them coins of different values. The animals quickly learned to judge the value of the coins and make the machines disgorge the bananas.

Under such conditions chimpanzees even learn to work for money. They also begin to beg for money and to steal it. But they never learn to regard the possession of money as a symbol of social status or rank. Consequently, the richer a chimpanzee is, the lazier he becomes. 'No vestige of capitalism would ever emerge in chimpanzee society.'

It is much the same with clothes and ornaments, with which captive chimpanzees sometimes like to deck themselves, or with drawing and painting, wherein these apes show incipient aesthetic feeling.[18] Everything they do is apparently done only for their own pleasure and remains without social consequences. A chimpanzee may happily don the scientist's white smock, but his cage mates will never feel for him the same awe and respect they have for the director of the laboratory. The symbolic value of uniform and insignia of rank is totally lost on a chimpanzee.

Under certain pressures such phenomena as theft, rape, blackmail, and prostitution have been noted among apes. But such temporary aberrations never lead to a lasting dependency relationship. Masters and slaves exist only among human beings. Chimpanzees may cheat one another, may go in for feints, histrionics, hypocrisy, evasions, and similar tricks which are familiar enough in human political life. Nonetheless these animals are far from practicing politics.

Capitalism, uniforms, slavery, and politics—these are matters known only to mankind, it would seem, and depend on a special capacity for setting up arrangements which outlive the individual. The 'arrangements' include all human customs, manners, statutes, morals, (in the plural) and technologies, everything from the idea of the state and the civil code to varieties of artistic style—in short, the entire infrastructure of our lives.

We can only guess at the biological factors which permitted that structure to arise. Dr. Kortlandt believes that five behavioral elements were involved:

1. Technological ability.
2. Interest in beautiful things.
3. The ability to make symbols.

These three factors are present in the ape, to be sure, but always exist in isolation. To these Dr. Kortlandt adds:

4. Foresight over long periods of time, which must have developed slowly when the ancestor of man changed from plant eater to hunter.
5. The need for cooperation and social stability when hunting became a community enterprise.

Out of the union of these characteristics there eventually arose the type of social organization in which the heritage of the past could be transmitted and carried on in noninstinctive fashion.

Ecce Homo sapiens.

Source Notes

Chapter 1

1 ADRIAAN KORTLANDT, *Some Results of a Pilot Study on Chimpanzee Ecology*. Obtainable from the Zoological Laboratory, University of Amsterdam, Plantage Doklaan 44, Amsterdam C, Holland.

2 ADRIAAN KORTLANDT, 'Chimpanzees in the Wild', *Scientific American*, Vol. 206, No. 5 (1962), pp. 128–38.

3 ADRIAAN KORTLANDT, *Enkele voorlopige resultaten van de Nederlandse chimpansee-expeditie*, 1963. Obtainable from the Zoological Laboratory, University of Amsterdam. *See* Note 1.

4 ADRIAAN KORTLANDT, 'Protohominid Behaviour in Primates', in *Symposia*, Vol. 10 (1963), Zoological Society of London, Regent's Park, London N.W.1, pp. 61–88.

5 ADRIAAN KORTLANDT, 'Bipedal Armed Fighting in Chimpanzees', in *Symposium on Behavior Adaptions to Environment in Mammals.* Washington, D.C., XVI International Congress of Zoology, 1963.

6 MAX GLUCKMAN, 'The Rise of a Zulu Empire', *Scientific American*, Vol. 202, No. 4 (1960), pp. 157–68.

7 WOLFGANG KÖHLER, *Intelligenzprüfungen an Menschenaffen*. Berlin, Springer, 1963.

8 ADRIAAN KORTLANDT, *Verslag derde chimpansee-expeditie*, 1964. Obtainable from Zoological Laboratory, University of Amsterdam. *See* Note 1.

9 ADRIAAN KORTLANDT, 'Some Experiments with Chimpanzees in the Wild in Order to Test the Dehumanisation Hypothesis on Ape Evolution', Manuscript.

10 *How Do Male Chimpanzees Use Weapons When Fighting with Leopards?* Yearbook, American Philosophical Society. In press.

[11] GAVIN DE BEER, *Atlas of Evolution*. London and Edinburgh, Thomas Nelson & Sons, 1964, pp. 174–7.

[12] R. A. DART, 'The Osteodontokeric Culture of Australopithecus Prometheus', Transvaal Museum Memorandum, No. 10 (1957).

[13] L. S. B. LEAKEY, 'Adventures in the Search of Man', *National Geographic Magazine*, Vol. 123 (1963), pp. 132–52.

[14] THEODOSIUS DOBZHANSKY, in *Mankind Evolving*. New York, Bantam Books, 1970, p. 186.

[15] ADRIAAN KORTLANDT, Personal communication to the author.

[16] R. A. BUTLER, 'Discrimination Learning by Rhesus Monkeys to Visual Exploration Motivation', *Journal of Comparative and Physiological Psychology*, Vol. 46 (1953), pp. 95–8.

[17] WOLFGANG KÖHLER, 'Zur Psychologie des Schimpansen', *Psychologische Forschungen*, Vol. 1 (1921), p. 16.

[18] IRVEN DEVORE, *Primate Behavior: Field Studies of Monkeys and Apes*. New York, Holt, Rinehart and Winston, 1965.

[19] JANE GOODALL, 'My Life Among Wild Chimpanzees', *National Geographic Magazine*, Vol. 124, No. 2 (1963), pp. 273–308.

[20] JANE VAN LAWICK-GOODALL, 'New Discoveries Among Africa's Chimpanzees', *National Geographic Magazine*, Vol. 128, No. 6 (1965), pp. 802–31.

[21] JANE GOODALL, 'Feeding Behaviour of Chimpanzees', in *Symposia*, Vol. 10 (1963). Zoological Society of London, Regent's Park, London N.W.1, pp. 39–47.

[22] JANE VAN LAWICK-GOODALL, *My Friends the Chimpanzees*. Washington, D.C., National Geographic Society, 1967.

[23] VINCENT M. SARICH and ALLAN C. WILSON, 'Immunological Time Scale for Hominid Evolution', *Science*, Vol. 158 (1967), pp. 1200–2.

Chapter 2

[1] JOHN C. LILLY, 'Distress Call of the Bottlenose Dolphin', *Science*, Vol. 139 (1963), pp. 116–18.

[2] JOHN C. LILLY, *Man and Dolphin*. Garden City, N.Y., Doubleday, 1961; London, Victor Gollancz, 1962.

[3] LEO SZILARD, *The Voice of the Dolphins and Other Stories*. New York, Simon and Schuster, 1961; London, Sphere Books, 1967.

[4] ANTONY ALPERS, *Dolphins, the Myth and the Mammal*. Boston, Houghton Mifflin, 1961.

[5] G. PILLERI, *Intelligenz und Gehirnentwicklung bei den Walen*. Basel, Sandoz-Panorama, 1962.

[6] WINTHROP N. KELLOGG, *Porpoises and Sonar*. Chicago, University of Chicago Press, 1961.

7 W. E. EVANS and J. H. PRESCOTT, 'Observations of the Sound Production Capabilities of the Bottlenose Porpoise', Zoologica N.Y., Vol. 47 (1962), pp. 121–8.

8 PETER STUBBS, 'Dolphin on the 'Phone', New Scientist, Vol. 29, No. 478 (1966), pp. 83–5, based on an original article in Science, Vol. 150, p. 1839.

9 For further details see Vitus B. Dröscher's The Mysterious Senses of Animals. New York, E. P. Dutton & Co., 1965; London, Hodder & Stoughton, 1965, p. 22.

10 JOHN C. LILLY, 'Vocal Mimicry in Tursiops: Ability to Match Numbers and Durations of Human Vocal Bursts', Science, Vol. 147 (1965), pp. 300–1.

11 R. G. BUSHNEL, 'Information in the Human Whistled Language and Sea Mammal Whistling', in K. Norris, Whales, Dolphins and Porpoises. Proceedings of the First International Symposium on Cetacean Research, Washington 1964. Berkeley, University of California Press, 1966, pp. 544–68.

12 Report in Christian Science Monitor, September 8, 1967.

13 KENNETH S. NORRIS, 'Trained Porpoise Released in the Open Sea', Science, Vol. 147, No. 3661 (1965), pp. 1048–50.

14 ERWIN TRETZEL, 'Imitation und Variation von Schäferpfiffen durch Haubenlerchen', Zeitschrift für Tierpsychologie, Vol. 22, No. 7 (1965), pp. 784 809.

15 J. TRIAR, 'Kurze Mitteilung ber üdie Haubenlerche des Herrn Kullman', Geflügelte Welt, Vol. 62 (1933), p. 358.

16 ERWIN TRETZEL, 'Imitation und Transposition menschlicher Pfiffe durch Amseln,' Zeitschrift für Tierpsychologie, Vol. 24, No. 2 (1967), pp. 137–61.

17 LORUS MILNE and MARGERY, The Senses of Animals and Men. New York, Atheneum, 1962, p. 69.

18 Article: 'Contrapuntal Bird Songs', Scientific American, Vol. 208, No. 5 (1963), pp. 80–1.

19 JOHANNES KNEUTGEN, Beobachtungen über die Anpassung von Verhaltensweisen an gleichförmige akustische Reize', Zeitschrift für Tierspsychologie, Vol. 21, No. 6 (1964), pp. 763–79.

20 KONRAD LORENZ, 'Die angeborenen Formen möglicher Erfahrung', Zeitschrifte für Tierspsychologie, No. 5 (1943), p. 235.

21 FRIEDRICH KAINZ, Die Sprache der Tiere. Stuttgart, Ferdinand Enke, 1961.

22 KONRAD LORENZ, Er redete mit dem Vieh, den Vögeln und den Fischen. Vienna, Borotha-Schoeler, 1949, pp. 49–106.

23 OTTO KOENIG, 'Das Aktionssystem der Bartmeise', Österreichische Zoologische Zeitschrift, Vol. 3 (1951), p. 247.

24 EDWARD O. WILSON, 'Pheromones', Scientific American, Vol. 208, No. 5 (1963), pp. 100–14.

25 MARTIN LINDAUER, Review of F. Kainz's book, *Die Sprache der Tiere*. *Naturwissenschaftliche Rundschau* 15, 9/10 (1962), pp. 412–13.

26 MARTIN LINDAUER, 'Die Sprache der Bienen'. Lecture at the Zoological Institute of Hamburg University, 1965.

27 MARTIN LINDAUER, 'Fortschritte der Zoologie', *Allgemeine Sinnesphysiologie*, Vol. 16, No. 1 (1963), p. 68.

28 EBERHARD GWINNER and JOHANNES KNEUTGEN, 'Über die biologische Bedeutung der "zweckdienlichen" Anwendung erlernter Laute bei Vögeln', *Zeitschrift für Tierpsychologie*, Vol. 19, No. 6 (1962), pp. 692–6.

29 EBERHARD GWINNER, 'Untersuchungen über das Ausdrucks- und Sozialverhalten des Kolkraben', *Zeitschrift für Tierpsycologie*, Vol. 21, No. 6 (1964), pp. 657–748, particularly pp. 690–700.

30 ERWIN TRETZEL, 'Über das Spotten der Singvögel', *Verhandlungen der Deutschen Zoologischen Gesellschaft*, Kiel (1964), pp. 556–65.

31 W. H. THORPE, *Learning and Instinct in Animals*. Cambridge, Harvard University Press, 1966; London, Methuen, 1966, pp. 16–17.

32 MASAKAZU KONISHI, 'The Role of Auditory Feedback in the Control of Vocalisation in the White-Crowned Sparrow', *Zeitschrift für Tierpsychologie*, Vol. 22, No. 7 (1965), pp. 770–83.

33 Report: 'Untutored Birds Sing Simpler Songs', *New Scientist*, Vol. 30 (1966), p. 313.

34 JOHANNES KNEUTGEN, 'Über die künstliche Auslösbarkeit des Gesangs der Schamadrossel', *Zeitschrift für Tierpsychologie*, Vol. 21, No. 1 (1964), pp. 124–8.

35 GERHARD THIELCKE, 'Vogel-laute und –gesänge', *Umschau in Wissenchaft und Technik*, Vol. 62 (1962), pp. 365–7.

36 RÉMY CHAUVIN, *Animal Societies, from the Bee to the Gorilla*. Translated [from the French] by George Ordish. London, Gollancz, 1968. New York, Hill and Wang, 1968, p. 190.

37 Report: *Deutscher Forschungsdienst*, March 1964.

38 JÜRGEN NICOLAI, 'Familientradition in der Gesangsentwicklung des Gimpels', *Journal für Ornithologie*, Vol. 100 (1959), pp. 39–46.

39 DETLEV PLOOG, 'Auf dem Weg zum denkenden Wesen', *BP-Kurier*, Vol. 2 (1964), p. 37.

40 FRIEDRICH KAINZ, *Die Sprache der Tiere*. Stuttgart, Ferdinand Enke (1961), pp. 141–8. (Examples of deception in the animal kingdom.)

41 ERICH BAEUMER, 'Das "dumme" Huhn', *Kosmos-Bibliothek*, Vol. 242. Stuttgart (1964), pp. 80–1.

42 *Ibid.*, p. 67.

43 ADOLF REMANE, 'Das soziale Leben der Tiere', *Rowohlts Deutsche Enzyklopädie*, Vol. 97. Hamburg, Rowohlt, 1960.

44 IVAN SANDERSON and FRITZ BOLLE, *Living Mammals of the World*. New York, Doubleday, 1961.

45 IRENÄUS EIBL-EIBESFELDT, *Land of a Thousand Atolls: A Study of Marine Life in the Maldive and Nicobar Islands*. Translated by Gwynne Vevers. London, McGibbon & Kee, 1965; Cleveland, The World Publishing Company, 1966, p. 74.

46 HEINI HEDIGER, *Beobachtungen zur Tierpsychologie im Zoo und im Zirkus*. Basel, Reinhardt, 1961, Chapter 7.

47 OTTO VON FRISCH, *Spaziergang mit Tobby*. Stuttgart, Franckh, 1963, pp. 50–1.

48 S. ZUCKERMANN, *The Social Life of Monkeys and Apes*. London, Kegan Paul, Trench, Trubner & Co ; New York, Harcourt, Brace & Company, 1932.

Chapter 3

1 J. PRÉVOST, *Ecologie du Manchot Empereur*. Paris, Hermann, 1961.

2 JEAN RIVOLIER, *Emperor Penguins*, translated by Peter Wiles. London, Elek Books, 1956; Toronto, Ryerson Press, 1956.

3 RÉMY CHAUVIN, *Animal Societies from the Bee to the Gorilla*. Translated [from the French] by George Ordish. London, Gollancz, 1968. New York, Hill and Wang, 1968, pp. 224 ff.

4 J. F. RICHDALE, *A Population Study of Penguins*. Oxford, Oxford University Press, 1957.

5 EBERHARD GWINNER, 'Untersuchungen über das Ausdrucks- and Sozialverhalten des Kolkraben', *Zeitschrift für Tierpsychologie*, Vol. 21, No. 6.

6 VITUS B. DRÖSCHER, 'Raben haben strenge Regeln', *Das Beste* (July 1967), pp. 54–60.

7 EBERHARD GWINNER, 'Über den Einfluss des Hungers und anderer Faktoren auf die Versteck-Aktivität des Kolkraben', *Die Vogelwarte*, Vol. 23, No. 1 (1965), pp. 1–4.

8 EBERHARD GWINNER, 'Beobachtungen über Nestbau und Bruptpflege des Kolkraben', *Journal für Ornithologie*, Vol. 106, No. 2 (1965), pp. 145–78.

9 HELGA FISCHER, 'Das Triumphgeschrei der Graugans', *Zeitschrift für Tierpsychologie*, Vol. 22, No. 3 (1965), pp. 247–304, particularly p. 300.

10 GUSTAV KRAMER, 'Beobachtungen und Fragen zur Biologie des Kolkraben', 'Journal für Ornithologie, Vol. 80 (1932), p. 329.

11 L. MOESGAARD, 'Ravnen som dansk ynglefugl', *Danske Fugle*, Vol. 9 (1929), p. 171.

12 EBERHARD GWINNER, 'Über einige Bewegungsspiele des Kolkraben', *Zeitschrift für Tierpsychologie*, Vol. 23, No. 1 (1966), pp. 28–36.

13 JOHANNES GOTHE, 'Zur Droh- und Beschwichtigungsgebärde des Kolkraben', *Zeitschrift für Tierpsychologie*, Vol. 19, No. 6 (1963), pp. 687–91.

[14] IRVEN DEVORE, *Primate Behavior: Field Studies of Monkeys and Apes.* New York, Holt, Rinehart and Winston, 1965.

[15] Report: 'Verhaltensbeobachtungen an frei lebenden Affen', *Umschau in Wissenschaft und Technik*, Vol. 65, No. 22 (1965), p. 717.

[16] M. KAWAI, 'Newly Acquired Pre-Cultural Behavior of the Natural Troop of Japanese Monkeys on Koshima Islet', *Primates*, Vol. 6 (1965), pp. 1–30.

[17] JOHN A. KING, 'The Social Behavior of Prairie Dogs', *Scientific American*, Vol. 201, No. 4 (1959), pp. 128–40.

[18] HUBERT and MABEL FRINGS, 'The Language of Crows', *Scientific American*, Vol. 201, No. 5 (1959), pp. 119–31.

[19] MARTIN LINDAUER, 'Schwarmbienen auf Wohnungssuche', *Zeitschrift für vergleichende Physiologie*, Vol. 37 (1955), pp. 263–324.

[20] S. G. CHEN, 'Social Modification of the Activity of Ants in Nest Building', *Physiological Zoology*, Vol. 10 (1937).

[21] T. C. SCHNEIRLA, 'The Behavior and Biology of Certain Nearctic Doryline Ants', *Zeitschrift für Tierpsychologie*, Vol. 18, No. 1 (1961), pp. 1–32.

[22] KARL GÖSSWALD, *Unsere Ameisen I*, *Kosmos-Bändchen* No. 204. Stuttgart, Franckh, 1954, p. 44.

[23] ROLF LANGE, 'Die Nahrungsverteilung unter den Arbeiterinnen des Waldameisenstaates', *Zeitschrift für Tierpsychologie*, Vol. 24, No. 5 (1967), pp. 513–45.

[24] U. MASCHWITZ, 'Parasitische Käferlarven imitieren die Brut ihrer Wirts-ameisen', *Naturwissenschaftliche Rundschau*, Vol. 21, No. 1 (1968), p. 26.

[25] HUBERT MARKL, 'Die Verständigung durch Stridulationssignale bei Blattschneiderameisen', *Zeitschrift für vergleichende Physiologie*, Vol. 57 (1967), pp. 299–330.

[26] ALEXANDER B. and ELSIE B. KLOTS, *Living Insects of the World.* Garden City, N.Y., Doubleday, 1959, p. 230.

[27] IRENÄUS and E. EIBL-EIBESFELDT, 'Das Parasitenabwehren der Minima-Arbeiterinnen der Blattschneiderameise', *Zeitschrift für Tierpsychologie*, Vol. 24, No. 3 (1967), pp. 278–81.

[28] Report: 'Elephants Tried to Move Their Dead.' *New Scientist*, Vol. 25, No. 428 (1965), p. 205.

[29] Report: 'Drama einer alten Elefantenkuh', *Naturwissenschaftliche Rundschau*, Vol. 19, No. 9 (1966), p. 382.

[30] NORMAN CARR, *Return to the Wild.* New York, E. P. Dutton & Co., 1962; London, Collins, 1962, p. 91.

[31] ERNA MOHR, 'Das Verhalten der Pinnipedier', *Handbuch der Zoologie*, Vol. 8 (1956).

[32] FRITZ DIETERLEN, 'Geburt und Geburtshilfe bei der Stachelmaus', *Zeitschrift für Tierpsychologie*, Vol. 19, No. 2 (1962), pp. 191–222.

33 BERNHARD GRZIMEK, 'Delphine helfen kranken Artgenossen', *Säugetierkundliche Mitteilungen*, Vol. 5 (1957), p. 160.
34 E. J. SLIJPER, 'Die Geburt der Säugetiere', *Handbuch der Zoologie*, Vol. 8, No. 9 (1960), pp. 1–108.
35 F. POPPLETON, 'Birth of an Elephant', *Oryx*, Vol. 4 (1957), p. 180.
36 W. C. OSMAN HILL, *Primates III*. Edinburgh, Edinburgh University Press, 1957.
37 A similar incident is reported by Fritz Walther, *Mit Horn und Huf*. Berlin, Paul Parey, 1966.
38 IVAN SANDERSON and FRITZ BOLLE, *Living Mammals of the World*. New York Doubleday, 1961, pp. 267–8.
39 KONRAD LORENZ, *On Aggression*. New York, Harcourt, Brace and World, 1967; London, Methuen, 1967, pp. 94–5.
40 E. GERSDORF, 'Beobachtungen über das Verhalten von Vogelschwärmen', *Zeitschrift für Tierpsychologie*, Vol. 23, No. 1 (1966), pp. 37–43.
41 KONRAD LORENZ, *Über tierisches und menschliches Verhalten*, Vol. 1. Munich, Piper Paperback, 1965, pp. 13–114, particularly pp. 19–23.
42 OTTO VON FRISCH, *Spaziergang mit Tobby*. Stuttgart, Franckh, 1963, pp. 91–110.
43 R. and R. MENZEL, 'Über Interferenzerscheinungen zwischen sozialer und biologischer Rangordnung', *Zeitschrift für Tierpsychologie*, Vol. 19, No. 3 (1962), pp. 332–55.

Chapter 4

1 V. C. WYNNE-EDWARDS, *Animal Dispersion in Relation to Social Behaviour*. Edinburgh, Oliver and Boyd, 1962.
2 V. C. WYNNE-EDWARDS, 'Population Control in Animals', *Scientific American*, Vol. 211, No. 2 (1964), pp. 68–74.
3 V. C. WYNNE-EDWARDS, 'Self-Regulating Systems in Populations of Animals', *Science*, Vol. 147 (1965), pp. 1543–8.
4 PETER H. KLOPFER, *Ökologie und Verhalten*. Stuttgart, Gustav Fisher Verlag, 1968, p. 33.
5 'Gray Seals Choose To Be Overcrowded', *New Scientist*, Vol. 24, No. 416 (1964), p. 342.
6 LORUS and MARGERY MILNE, *The Balance of Nature*. New York, Alfred A. Knopf, 1960; London, André Deutsch, 1961.
7 DAVID LACK, 'Are Bird Populations Regulated?' *New Scientist*, Vol. 31, No. 504 (1966), pp. 98–9.
8 Report: 'Drought Stops the Rabbit Breeding', *New Scientist*, Vol. 24, No. 416 (1964), p. 386.

[9] JOHN B. CALHOUN, 'Population Density and Social Pathology', *Scientific American*, Vol. 206, No. 2 (1962), pp. 139–48.

[10] OTTO KOENIG, 'Wohlstandsverwahrlosung in einer Kuhreiherkolonie', *Journal für Ornithologie*, Vol. 107, No. 3/4 (1966), pp. 406–7.

[11] Report: 'A Smell Makes Lone Mice Feel Crowded', *New Scientist*, Vol. 31, No. 510 (1966), pp. 430–1.

[12] H. M. BRUCE, 'Time Relations in the Pregnancy-Block Induced in Mice by Strange Males', *Journal of Reproduction and Fertilisation*, Vol. 2 (1961), p. 138.

[13] DETLEV PLOOG, 'Auf dem Weg zum denkenden Wesen', *BP–Kurier*, Vol. 2 (1964), p. 37.

[14] T. T. Macan, 'Self-Controls on Population Size', *New Scientist*, Vol. 28, No. 474 (1965), pp. 801–3.

[15] BERNHARD GRZIMEK, *Serengeti Shall Not Die*. New York, E. P. Dutton & Co., 1962.

[16] Report: 'Elephant Invasion in the Serengeti', *New Scientist*, Vol. 36, No. 576 (1967), p. 704.

[17] Report: 'Family Planning Among Elephants', *New Scientist*, Vol. 32, No. 519 (1966), p. 215.

[18] Report: 'Giant Tortoises Seem Unwilling to Breed', *New Scientist*, Vol. 36, No. 573 (1967), p. 528.

[19] T. G. SCHULTZE-WESTRUM, 'Biologische Grundlagen zur Populationsphysiologie der Wirbeltiere', *Die Naturwissenschaften*, Vol. 54, No. 22 (1967), pp. 576–9.

[20] PAUL LEYHAUSEN, 'Gesunde Gemeinschaft—ein Dichteproblem?' *BP–Kurier*, Vol. 2/3 (1966), pp. 21–7.

[21] CORINNE HUTT and JANE VAISEY, 'Personality and Overcrowding', *Nature*, Vol. 209 (1966), p. 1371.

[22] S. L. WASHBURN and IRVEN DeVORE, 'The Social Life of Baboons', *Scientific American*, Vol. 204, No. 6 (1961), pp. 62–71.

[23] K. POECK, 'Hypochondrische Entwurzelungsdepressionen bei italienischen Arbeitern in Deutschland, *Umschau in Wissenschaft und Technik*, Vol. 63 (1963), p. 354.

[24] GEORGE M. CARSTAIRS, at the Eighth Conference of the International Society for Planned Parenthood on November 4, 1967, at Santiago, Chile.

[25] RÉMY CHAUVIN, *Animal Societies, from the Bee to the Gorilla*. Translated [from the French] by George Ordish. London, Gollancz, 1968. New York, Hill and Wang, 1968, p. 150.

[26] C. B. WILLIAMS, *Insect Migration*. London, Collins, 1958, p. 85.

[27] Report: 'What Drives the Lemmings On?' *New Scientist*, Vol. 26, No. 437 (1965), p. 10.

28 H. U. THIELE, 'Neue Beobachtungen zum Rätsel der Lemmingwanderungen', *Naturwissenschaftliche Rundschau*, Vol. 18, No. 4 (1965), pp. 156–7.

29 Report: 'Built-In Armour for Hot-Headed Lemmings', *New Scientist*, Vol. 27, No. 461 (1965), p. 698.

30 ERICH BAEUMER, 'Verhaltensstudie über das Haushuhn', *Zeitschrift für Tierpsychologie*, Vol. 16 (1959), pp. 284–96.

31 ERIC BERNE, *Games People Play*. New York, Grove Press, 1964.

32 PAUL LEYHAUSEN, 'Zur Naturgeschichte der Angst', *Politische Psychologie*, Vol. 6. Frankfurt (Main), Europaïsche Verlagsanstalt (1967), pp. 94–112.

33 KONRAD LORENZ, *On Aggression*. New York, Harcourt, Brace and World, 1967; London, Methuen, 1967.

34 IRENÄUS EIBL-EIBESFELDT, *Grundriss der vergleichenden Verhaltensforschung*. Munich, Piper, 1967, p. 284.

35 KONRAD LORENZ, in *Aspekte der Angst*, edited by Hoimar von Ditfurth. Stuttgart, Georg Thieme, 1965, p. 19.

36 *Ibid.*, p. 16.

37 SHELDON and ELEANOR GLUECK, *A Manual of Procedures for Application of the Glueck Prediction Table*. New York, Youth Board Research Institute of New York.

38 See Note 35, p. 19.

39 GRAHAM CHEDD, 'A Cause for Anxiety', *New Scientist*, Vol. 37, No. 581 (1968), pp. 183–4.

Chapter 5

1 NORMAN CARR, *Return to the Wild*. New York, E. P. Dutton & Co., 1962; London, Collins, 1962.

2 RUDOLF SCHENKEL, 'Zum Problem der Territorialität', *Zeitschrift für Tierpsychologie*, Vol. 23, No. 5 (1966), pp. 593–626.

3 RUDOLF SCHENKEL, '*Über das Sozialleben der Löwen in Freiheit*, Vol. 12. Basel, Zolli (1964), p. 14.

4 BERNHARD GRZIMEK, *Serengeti Shall Not Die*. London, Hamish Hamilton, 1960; New York, E. P. Dutton & Co., 1961, p. 81.

5 VITUS B. DRÖSCHER, *The Mysterious Senses of Animals*. London, Hodder & Stoughton, 1965; New York, E. P. Dutton & Co., pp. 59–67.

6 RUDOLF SCHENKEL, 'Toten Lowen ihre Artgenossen?' *Umschau in Wissenschaft und Technik*, Vol. 68, No. 6 (1968), pp. 172–4.

7 GEORGE B. RABB, 'How Wolves Become Friends', *Science Service*, August 24, 1966.

8 HANS KRUUK, 'A New View of the Hyaena', *New Scientist*, Vol. 30, No. 502 (1966), pp. 849–51.

9 See Chapter 4, Note 33, pp. 241–54.

[10] S. A. Barnett, 'Rats', *Scientific American*, Vol. 216, No. 1 (1967) pp. 78–85.

[11] Report: 'Adult Rats Emit Ultrasounds', *New Scientist*, Vol. 35, No. 557 (1967), p. 281.

[12] H. Oldfield-Box, 'Social Organisation of Rats in a "Social Problem" Situation', *Nature*, Vol. 213, No. 5075 (1967), pp. 533–4.

Chapter 6

[1] Roger Ulrich, 'Pain as a Cause of Aggression', Berkeley Meeting of the American Association for the Advancement of Science, 1965.

[2] Ulrich, Stachnik, Brierton, and Mabry, 'Fighting and Avoidance in Response to Aversive Stimulation', *Behaviour*, Vol. 26 (1966), pp. 124–9.

[3] Walter Vernon and Roger Ulrich, 'Classical Conditioning of Pain-Elicited Aggression', *Science*, Vol. 152 (1966), p. 668.

[4] Walter C. Rothenbuhler, 'Behavior Genetics of Nest Cleaning in Honeybees', *American Zoology*, Vol. 4 (1964), pp. 111–23.

[5] Erich von Holst, 'Vom Wirkungsgefüge der Triebe', *Die Naturwissenschaften*, Vol. 18 (1960), pp. 409–22.

[6] 'Missing Enzyme Can Cause Aggression', *New Scientist*, Vol. 34, No. 540 (1967), p. 75.

[7] Oliver la Farge, *A Pictorial History of the American Indian*. New York, Crown Publishers, 1956; London, André Deutsch, 1958.

[8] Clyde Edgar Keeler, *Land of the Moon-Children*. Chicago, University of Chicago Press, 1960.

[9] John P. Scott, *Aggression*. Chicago, University of Chicago Press, 1960.

[10] Z. Y. Kuo, 'The Genesis of the Cat's Responses to the Rat', *Journal of Comparative Psychology*, Vol. 11 (1960), pp. 1–35.

[11] J. Dollard, et al, *Frustration and Aggression*. New Haven, Yale University Press, 1939.

[12] Eugen Gürster, 'Der schwierige Umgang mit dem "sogenannten Bösen" ', *Die Welt der Literatur*, January 1, 1965, p. 15.

[13] Jean-François Steiner, *Treblinka*, translated by Helen Weaver. New York, Simon & Schuster, 1967; London, Weidenfeld & Nicolson, 1967.

[14] Stanley Milgram, 'Behavioral Study of Obedience', *Journal of Abnormal Social Psychology*, Vol. 67 (1963), pp. 372–8.

[15] Stanley Milgram, 'Einige Bedingungen von Autoritätsgehorsam und seiner Verweigerung', *Zeitschrift für experimentelle und angewandte Psychologie*, Vol. 13 (1966), pp. 433–63.

[16] Irenäus Eibl-Eibesfeldt, *Grundriss der vergleichenden Verhaltensforschung* Munich, Piper (1967), pp. 437–8.

[17] Report: 'Taming Ferrets with Food', *New Scientist*, Vol. 22, No. 389 (1964), p. 293.

[18] J. K. KOVACH, in *Science*, Vol. 156 (1967), p. 835.
[19] JENS BJERRE, *Kalahari*. Translated from the Danish by Estrid Bannister. New York, Hill & Wang, 1960; London, Michael Joseph, 1960.

Chapter 7

[1] DETLEV PLOOG, *The Behavior of Squirrel Monkeys*. Chicago, Chicago University Press, 1965.

[2] DETLEV PLOOG, 'Vergleichend quantitative Verhaltensstudien an zwei Totenkopffaffen-Kolonien, *Zeitschrift für morphologische Anthropologie*, Vol. 53 (1963), pp. 92–108. A personal account.

[3] PETER WINTER, 'Verständigung durch Laute bei Totenkopffaffen', *Umschau*, Vol. 66, No. 20 (1966), pp. 653–8.

[4] J. ITANI, *et al.*, 'The Social Construction of Natural Troops of Japanese Monkeys in Takasakiyama', *Primates*, Vol. 4 (1963), pp. 1–42.

[5] M. KAWAI, *Ecology of Japanese Monkeys*. Tokyo, Kawadeshoboshinsha, 1964.

[6] LORUS and MARGERY MILNE, *The Senses of Animals and Men*. London, André Deutsch, 1963; New York, Atheneum, 1962, p. 47.

[7] R. DEAN and MARCELLE GEBER, 'The Development of the African Child', *Discovery* (1964), pp. 14–19.

[8] HARRY F. HARLOW, 'The Nature of Love', *American Psychologist*, Vol. 12, No. 13 (1958), pp. 673–85.

[9] HARRY F. HARLOW, 'Love in Infant Monkeys', *Scientific American*, Vol. 200, No. 6 (1959), pp. 68–74.

[10] HARRY F. HARLOW, 'Social Deprivation in Monkeys', *Scientific American*, Vol. 207, No. 5 (1962), pp. 136–46.

[11] WOLFGANG WICKLER. '"Erfindungen" und die Entstehung von Traditionen bei Affen', *Umschau in Wissenschaft und Technik*, Vol. 67, No. 22 (1967), pp. 725–30.

[12] SEYMOUR LEVINE, 'Stimulation in Infancy', *Scientific American*, Vol. 202, No. 5 (1960), pp. 81–6.

[13] VICTOR H. DENENBERG, 'Early Experience and Emotional Development', *Scientific American*, Vol. 208, No. 6 (1963), pp. 138–46.

[14] C. KAUFMAN and L. A. ROSENBLUM, 'The Effects of Brief Mother–Infant Separation in Monkeys.' Downstate Medical Center Report, 1966.

[15] Report: 'Wholesome Pain', *Science Service* (1966).

[16] R. A. HINDE, 'The Effects of Maternal Deprivation in Monkeys', *Nature*, Vol. 210 (1966), p. 1021.

Chapter 8

[1] ALEXANDER B. and ELSIE B. KLOTS. *Living Insects of the World*. Garden City, N.Y.; Doubleday, 1959, p. 230.

2 *Ibid.*, p. 280.

3 ADOLF REMANE, *Das soziale Leben der Tiere.* Hamburg, Rowohlt (1960). p. 73.

4 ERICH HUTH, ' "Einemsen" beim Buchfinken', *Journal für Ornithologie*, Vol. 92, No. 1 (1951), p. 62–3.

5 R. GOTTSCHALK, 'Beobachtungen über das Einemsen', *Gefiederte Welt*, Vol. 8 (1966), pp. 157–8.

6 E. THOMAS GILLIARD, *Living Birds of the World.* London, Hamish Hamilton, 1958; Garden City, N.Y.; Doubleday, 1958, p. 218.

7 IRENÄUS EIBL-EIBESFELDT, 'Über den Werkzeuggebrauch des Spechtfinken', *Zeitschrift für Tierpsychologie*, Vol. 18, No. 3 (1961), pp. 343–6.

8 E. CURIO and P. KRAMER, 'Vom Mangrovefinken', *Zeitschrift für Tierpsychologie*, Vol. 21, No. 2 (1964), pp. 223–34.

9 JANE and HUGO VAN LAWICK-GOODALL, 'Use of Tools by the Egyptian Vulture', *Nature*, Vol. 212, No. 5069 (1966), pp. 1468–9.

10 ERWIN STRESEMANN, 'Der australische Bussard zertrümmert Eier durch Steinwurf', *Journal für Ornithologie*, Vol. 96 (1955), p. 215.

11 K. R. L. HALL and GEORGE B. SCHALLER, ' "Sea Otters" Tools for Opening Mussels', *Journal of Mammalology*, Vol. 45, No. 2 (1964), p. 287.

12 LARS WILSSON, *Biber—Leben und Verhalten.* Wiesbaden, F. A. Brockhaus, 1966.

13 HENRI J. HOFFMANN, 'Der Biber und das Bibergeil', *Dragoco Report*, Vol. 11 (1964), pp. 247–55.

14 LEONARD LEE RUE, III, *The World of the Beavers.* Folkestone, Bailey Brothers & Swinfen, 1965; New York, J. B. Lippincott, 1964, pp. 59–60.

15 MASAO KAWAI, on the red-faced macaque in *Grzimeks Tierleben*, Vol. 10. Munich, Kindler Verlag, 1967, pp. 431–2.

16 ALISON JOLLY, 'Lemur Social Behavior and Primate Intelligence', *Science*, Vol. 153, pp. 501–6.

17 DESMOND MORRIS, *The Naked Ape.* New York, Dell, 1969; London, Jonathan Cape, 1967.

18 THORSTEN KAPUNE, 'Untersuchungen zur Bildung eines "Wertbegriffs" bei niederen Primaten', *Zeitschrift für Tierpsychologie*, Vol. 23, No. 3 (1966), pp. 324–63.

19 M. E. BITTERMAN, "The Evolution of Intelligence', *Scientific American*, Vol. 212, No. 1 (1965), pp. 92–100.

20 ALBERT BANDURA, 'Behavioral Psychotherapy', *Scientific American*, Vol. 216, No. 3 (1967), pp. 78–86.

Chapter 9

1 N. BOLWIG, 'Facial Expression in Primates', *Behaviour*, Vol. 22 (1964), pp. 167–92.

[2] DESMOND MORRIS, *Von Wölfen und Hunden.* Sandoz-Panorama, 1963.

[3] RICHARD J. ANDREW, 'Die Evolution von Gesichtsausdrücken', *Umschau in Wissenschaft und Technik*, Vol. 68, No. 3 (1968), pp. 75–8.

[4] RICHARD J. ANDREW, 'The Origin and Evolution of the Calls and Facial Expressions of the Primates', *Behaviour*, Vol. 20 (1963), pp. 1–109.

[5] FRIEDRICH KAINZ, *Die Sprache der Tiere.* Stuttgart, Ferdinand Enke, 1961, p. 140.

[6] ADRIAAN KORTLANDT, 'On the Essential Morphological Basis for Human Culture', *Current Anthropology*, Vol. 6 (1965), pp. 320–26.

[7] C. HAYES, *The Ape in Our House.* London, Victor Gollancz, 1952.

[8] See Note 5, p. 162.

[9] R. A. and B. T. GARDNER, *Acquisition of Sign Language in the Chimpanzee.* Reno, University of Nevada Progress Report, 1967.

[10] OTTO JESPERSEN, *Language, Its Nature, Development and Origin.* New York, Henry Holt & Co., 1922.

[11] M. MIYADI, 'Social Life of Japanese Monkeys,' *Science*, Vol. 143, No. 3608 (1964), pp. 783–6.

[12] RICHARD J. ANDREW, 'Trends Apparent in the Evolution of Vocalisation in the Old World Monkeys and Apes', in *The Primates, 1962: Symposia*, Vol. 10 (1963), Zoological Society of London, Regent's Park, London N.W.1, pp. 89–102.

[13] R. W. SPERRY, 'Cerebral Organisation and Behavior', *Science*, Vol. 133, No. 3466 (1961), pp. 1749–57.

[14] BOGEN FISHER and VOGEL, 'Cerebral Commissurotomy', *Journal of the American Medical Association*, Vol. 194, No. 12 (1965), pp. 1328–9.

[15] MICHAEL S. GAZZANIGA, 'The Split Brain in Man', *Scientific American*, Vol. 217, No. 2 (1967), pp. 24–9.

[16] VITUS B. DRÖSCHER, *The Magic of the Senses.* New York, E. P. Dutton & Co., 1969; London, W. H. Allen & Co. Ltd., 1969, p. 11.

[17] KONRAD LORENZ, *Man Meets Dog.* New York, Penguin, 1965.

[18] DESMOND MORRIS, *The Biology of Art.* New York, Alfred A. Knopf, 1962; London, Methuen, 1962.

Subject Index

Author Index